Sustainability Sciences in Asia and Africa

Sustainable Agriculture and Food Security

Sustainability as a scientific endeavor requires broad understanding of the interconnections in our global environment. It is therefore positioned at the crossroads of research in natural sciences, technology and human sciences. This book series supports the global efforts by providing timely coverage of the progress, opportunities, and challenges towards achieving sustainable development goals (SDGs) in Asia and Africa. It fosters the research in transdisciplinary academic fields spanning across agriculture and global food security, health and diseases, access to clean water and affordable energy, climate resilience and effects of global climate change on biodiversity, water and land. This series is planned to narrate the stories of scientific development and progression towards SDGs from the regions of Asia and Africa. The overall aim of the series is to provide a comprehensive publication platform and act as a knowledge engine in the growth of sustainability sciences with a special focus on developing nations. The series constitute of four sub-series each focusing on a specific theme closely linked to a set of SDGs. Each sub-series is leaders and experts in agricultural sciences, environmental sciences, biomedicine and medical sciences respectively. Books in this series are useful for environmentalists, biotechnologists, plant scientists, medical practitioners, conservation biologists, policy makers, NGOs and scientific bodies working in the areas of sustainability sciences in the developing regions.

This book series support the global efforts towards sustainability by providing timely coverage of the progress, opportunities, and challenges of sustainable food production and consumption in Asia and Africa. The series narrates the success stories and research endeavors from the regions of Africa and Asia on issues relating to SDG 2: Zero hunger. It fosters the research in transdisciplinary academic fields spanning across sustainable agriculture systems and practices, post- harvest and food supply chains. It will also focus on breeding programs for resilient crops, efficiency in crop cycle, various factors of food security, as well as improving nutrition and curbing hunger and malnutrition. The focus of the series is to provide a comprehensive publication platform and act as a knowledge engine in the growth of sustainability sciences with a special focus on developing nations. The series will publish mainly edited volumes but some authored volumes. These volumes will have chapters from eminent personalities in their area of research from different parts of the world.

Santosh Kale • Rakesh S
Pushpajeet Choudhari
Gajanan Sawargaonkar • Ramesh Singh
Mangi Lal Jat

Soil Carbon Fractionation

Techniques and Protocols

Springer

ICRISAT
INTERNATIONAL CROPS RESEARCH
INSTITUTE FOR THE SEMI-ARID TROPICS

Santosh Kale
International Crops Research Institute
for the Semi-Arid Tropics
Hyderabad, Telangana, India

Pushpajeet Choudhari
International Crops Research Institute
for the Semi-Arid Tropics
Hyderabad, Telangana, India

Ramesh Singh
International Crops Research Institute
for the Semi-Arid Tropics
Hyderabad, Telangana, India

Rakesh S
International Crops Research Institute
for the Semi-Arid Tropics
Hyderabad, Telangana, India

Gajanan Sawargaonkar
International Crops Research Institute
for the Semi-Arid Tropics
Hyderabad, Telangana, India

Mangi Lal Jat
Department of Agricultural Research
and Education (DARE)
The Indian Council of Agricultural
Research (ICAR)
New Delhi, Delhi, India

ISSN 2730-6771 ISSN 2730-678X (electronic)
Sustainability Sciences in Asia and Africa
ISSN 2730-6798 ISSN 2730-6801 (electronic)
Sustainable Agriculture and Food Security
ISBN 978-981-95-4067-9 ISBN 978-981-95-4068-6 (eBook)
https://doi.org/10.1007/978-981-95-4068-6

This Springer imprint is published by the registered company Springer Nature Singapore Pte Ltd.
The registered company address is: 152 Beach Road, #21-01/04 Gateway East, Singapore 189721, Singapore

Foreword

As the world faces mounting challenges such as soil degradation, climate change, and food insecurity, the importance of soil carbon has never been greater. Soil organic carbon (SOC) is fundamental to sustaining soil health, supporting vital ecosystem services, and mitigating the impacts of climate change. Accurate measurement and characterization of SOC fractions are essential for understanding carbon dynamics, developing effective soil management strategies, and promoting sustainable land use practices.

In response to the growing need for standardized methods and protocols, we are pleased to present this comprehensive Manual on SOC Fractionation Protocols in Soils. This handbook provides a detailed guide to the principles, methodologies, and applications of carbon fractionation. It covers a range of techniques for characterizing SOC fractions, including active, slow, and passive carbon pools, as well as the assessment of key soil enzymes.

Developed by a team of scientists at ICRISAT, this manual is designed to meet the needs of students, researchers, and professionals in soil science, agriculture, ecology, and environmental management. By offering a standardized framework for SOC fractionation, it aims to enhance comparability and integration of data across studies and regions, contributing to a deeper understanding of soil carbon dynamics.

I am confident that this publication will serve as a valuable resource in advancing research and practice in soil carbon management, fostering sustainable land use, and strengthening climate resilience. I congratulate the Resilient Farming Systems (RFS) team at ICRISAT for this remarkable initiative in support of the global carbon research community.

Director General, International Crops Research Institute for the Semi-Arid Tropics
Hyderabad, India

Himanshu Pathak

Preface

Soil health has become increasingly recognized in recent years as a crucial aspect not only for agriculture but also for environmental sustainability and climate resilience efforts worldwide. At the core of soil health is the notion of soil carbon, which is an element that impacts soil fertility and water retention while playing a significant role in the global carbon cycle. The significance of preserving and enhancing soil carbon levels for the sake of soil health and CO_2 reduction is capturing the attention of various stakeholders, such as policymakers' organizations and land stewards. To effectively encourage carbon sequestration practices in cultivated lands, it is crucial to have affordable methods to measure changes in soil carbon stocks and predict how soil carbon will react to various management techniques, weather patterns, and soil conditions. Despite its importance, there is a lack of easily accessible and comprehensive resources dedicated to analyzing soil carbon which plays a key role in the voluntary carbon markets that have the potential to incentivize small to marginal farmers through sequestering verifiable amounts of soil carbon and improve farm resiliency in a changing climate.

Thus, the development of a Soil Carbon Analysis Manual stems from the pressing need to bridge this knowledge gap. As agricultural practices evolve and climate challenges intensify, the demand for practical, scientifically grounded tools to assess and enhance soil carbon levels has never been greater. This manual aims to meet that demand by providing a user-friendly, yet technically robust, resource that guides users through the intricacies of soil carbon measurement, interpretation, and management. The photometric and spectroscopic methods are included as they generate smaller volumes of toxic waste and can be a potential method to substitute the traditional wet digestion methods like Walkley and Black method, which in many countries is still widely used in routine soil carbon analysis.

In developing this manual, we have drawn upon the expertise of ICRISAT's soil scientists, agronomists, and analysts, ensuring that the content is both scientifically rigorous and accessible. This Soil Carbon Analysis Manual is designed to serve as a practical guide for a wide range of users, from researchers, analysts, and agronomists to policymakers. It seeks to synthesize the standard operating procedures for quantifying the various carbon pools using vivid protocols, offering clear, step-by-step instructions, supported by scientific insights and literature from peer-reviewed journals. This manual aims to foster more informed decision-making, ultimately contributing to healthier soils, more resilient agricultural systems, and a more sustainable future by empowering users with the knowledge and tools they need. We hope that this resource will not only fill a critical knowledge gap but also inspire further exploration and innovation in the field of soil carbon management.

Hyderabad, Telangana, India — Santosh Kale
Hyderabad, Telangana, India — Rakesh S
Hyderabad, Telangana, India — Pushpajeet Choudhari
Hyderabad, Telangana, India — Gajanan Sawargaonkar
Hyderabad, Telangana, India — Ramesh Singh
New Delhi, Delhi, India — Mangi Lal Jat

Preface to the Book Series

The UN's second Sustainable Development Goal (SDG) sets out to end hunger, achieve food security and improved nutrition, and promote sustainable agriculture by 2030. Yet despite the tireless efforts of the farmers, researchers, and policymakers the world over, over 700 million people still suffer from undernourishment. To address this food security gap, global food production systems will need to expand their capacities and produce almost twice their existing levels to feed the world's ever-growing population.

This poses a significant challenge, particularly in relation to climate change. Climate change negatively impacts agricultural productivity through numerous mechanisms, including heat stress on crops, unpredictable and extreme weather events that result in floods or droughts, significant shifts to growing seasons, and heightened pest pressures. Traditionally, we have made up for lost productivity by clearing forests and converting grasslands for agriculture. However, through deforestation and land conversion, we're worsening the effects of climate change by releasing stored carbon and eliminating carbon sinks.

It is therefore essential that we break free of this cycle by establishing pathways toward sustainable crop production that enable us to address global food insecurity without exacerbating one of the key drivers of low agricultural productivity. However, it is also vital that efforts to deliver sustainable global food systems do not come at the detriment of farmers, and that they factor in the everyday realities of participants from across the agriculture value chain.

In recent years, the adoption of new research approaches has resulted in improvements in food access, farmer autonomy, and the stabilization of agriculture's global carbon footprint. Yet, the challenge to reduce hunger remains pronounced in the Global South, particularly Asia and Africa, which contains the largest proportion of people suffering from malnutrition and other nutrition-related issues. Furthermore, the agri-food systems in Asia and Africa are severely constrained by the subsistence nature of farming, barriers to accessing new research tools and technologies, and the increasing impacts of environmental pollution, soil and biodiversity degradation.

This book series, Sustainable Agriculture and Food Security, aims to support global efforts toward sustainability by providing timely coverage of the progress, opportunities, and challenges of sustainable food production and consumption in

Asia and Africa. The series narrates the success stories and research endeavors from the regions of Africa and Asia on issues relating to SDG 2: Zero Hunger. It fosters research in transdisciplinary academic fields spanning sustainable agriculture systems and practices, pre-seed and post-harvest innovations, and food supply chains. The focus of the series is to provide a comprehensive publication platform and to act as a repository of knowledge in sustainability sciences with a particular focus on developing nations.

This book, *Soil Carbon Fractionation: Techniques and Protocols*, edited by Santosh Kale, Rakesh S, Pushpajeet Choudhari, Gajanan Sawargaonkar, Ramesh Singh, and Mangi Lal Jat focuses on an essential research discipline for the advancement of sustainable agriculture. This analytical technique enables us to see how different farming practices affect soil carbon, and helps us to better estimate how long carbon will remain sequestered. Importantly, soil carbon fractionation can help us find a balance between productivity and carbon sequestration, ensuring that long-term climate benefits can be achieved without sacrificing yield, increasing the likelihood of farmers adopting alternative farming practices.

The book presents 18 chapters written by experts and practitioners from across the globe, each containing up-to-date and important insights into this important topic. I commend the Editors for including multidisciplinary perspectives and would like to congratulate all the authors for their valuable contributions.

I wish to extend my sincere thanks and gratitude to the Springer staff, particularly Sanchi Bhimrajika, Associate Editor (Books, Medicine and Life Sciences), and Nick Melchior, Editorial Director (Books, Medicine and Life Sciences, Asia), for their constant support in the accomplishment of this compendium. The cooperation received from my supervisors and research leaders, such as Prof. Andrew Deeks (Vice Chancellor), Prof. Peter Eastwood (Deputy Vice Chancellor- Research), Prof. Peter Davies (Pro Vice Chancellor), and my laboratory colleagues—Rhys G. R. Copeland, Anu Chitikineni, and Vanika Garg from Murdoch University (Australia), are also gratefully acknowledged. I would like to thank my family members—Monika Varshney, Prakhar Varshney, and Preksha Varshney, for their love and support in discharging my duties as Series Editor.

Perth, WA, Australia Rajeev K. Varshney

Acknowledgments

The technical Handbook entitled *Soil Carbon Fractionation—Techniques and Protocols: A Handbook of Laboratory Protocols* was published by the, ICRISAT's Global Research Program –Resilient Farming Systems (RFS), International Crops Research Institute for the Semi-Arid Tropics, Patancheru, Hyderabad, India-502324.

We extend our sincere thanks and gratitude to

1. Indian Council of Agricultural Research (ICAR), New Delhi, India
2. Mahatma Phule Krishi Vidyapeeth (MPKV), Rahuri, Maharashtra, India

We also sincerely acknowledge the valuable support and assistance provided by Ms. Chetna Nimje and Ms. Bhavaniprasanna Kesanakurthi during the preparation of this handbook.

We also owe gratitude to everyone who contributed, either directly or indirectly, to the publication of this handbook. We genuinely believe that this book will be of great use to research scholars, institute-level scientists, officials, and policymakers engaged in carbon sequestration in relation to sustainable farming, nutrient management, and climate mitigation.

Competing Interests The authors have no competing interests to declare that are relevant to the content of this manuscript.

Executive Summary

Soil organic carbon (SOC) fractionation helps to ascertain how different carbon fractions/pools change their state and rate over time in various soil systems. Compared to total soil organic carbon, the carbon fractions with a substantially higher turnover rate can quickly and sensitively indicate the dynamics of soil quality.

This manual on the SOC fractionation protocol provides a thorough and systematic understanding of carbon fractions and their estimation laboratory procedures in the very simplest way. This handbook is designed exclusively for soil carbon researchers, viz. research scholars, scientists, and academicians, with clear-cut information that makes readers grasp the intricate ideas of C fractionation with ease. The present manual consolidates all procedures related to SOC fractionation, i.e., active-, slow-, and passive-carbon pools into one document, with detailed methodologies for determining SOC fractions. Each procedure is supported by appropriate chemical reactions, principles, step-by-step protocol as flowcharts, and calculations for a thorough understanding and correct application. Furthermore, to help the readers practically comprehend the experimental work, we have included some real images taken during the laboratory analysis, endpoints, or equipment used for the specific protocol.

No doubt, this handbook is extremely helpful for all soil carbon researchers who are interested in exploring the changes in soil carbon dynamics under agricultural management in various soil systems. We believe that it is a trusted resource for accurately assessing carbon dynamics in soils as there are no such handbooks like this that have simplified the procedures of different soil carbon pools. Simplifying these complex processes into an easily accessible form bridges the gap between theoretical knowledge and practical implementation, contributing to a better understanding of soil health and its role in carbon sequestration.

Contents

1 **Background** 1
1.1 Carbon Fractions in Relation to Carbon Farming 2
1.2 Carbon Fractions in Relation to Nutrient Cycling 4
1.3 Soil Carbon Sequestration and Climate Change Mitigation 5
1.4 Basic Principles in Analytical Chemistry 6
1.5 Soil Sampling and Processing 13
1.6 Laboratory Safety and Handling Practices 15
Suggested Reading 16

2 **Permanganate Oxidizable Carbon (POXC)** 19
2.1 Introduction 19
2.2 Principle 20
2.3 Instrumentation and Materials 20
2.4 Sample Reaction 22
2.5 Sample Dilution 22
2.6 Reading Samples on Spectrophotometer 22
2.7 Calculating Mass of POXC for Unknown Soil Samples 24
2.8 Example Calculation 24
2.9 Clean-Up and Disposal 24
Suggested Reading 24

3 **Water-Soluble Carbon in Soils (WSC)** 25
3.1 Introduction 25
3.2 Principle 26
3.3 Reagents/Materials Required 26
3.4 Procedure 26
3.5 Calculations (Fig. 3.1) 28
3.6 Precautions 28
Suggested Reading 28

4 **Soil Microbial Biomass Carbon (SMBC)** 29
4.1 Introduction 29
4.2 Principle 29
4.3 Reagents/Materials 30
4.4 Equipment/Instruments 30

4.5 Procedure 30
4.6 Calculations 31
4.7 Precautions 31
Suggested Reading 33

5 Light Fraction (LF) Carbon 35
5.1 Introduction 35
5.2 Principle 36
5.3 Reagents/Materials 36
5.4 Procedure 36
5.5 Calculations 38
5.6 Precautions 38
Suggested Reading 38

6 Determination of Soil Acid Hydrolysable Carbohydrates 39
6.1 Introduction 39
6.2 Principle 39
6.3 Reagents 40
6.4 Procedures 40
6.5 Calculation 42
Suggested Reading 42

7 Labile Pool-I and Labile Pool-II Carbon: Two-Step Acid Hydrolysis 43
7.1 Introduction 43
7.2 Principle 44
7.3 Procedure 44
7.4 Equipment 45
7.5 Reagents 45
7.6 Calculation 47
Suggested Reading 47

8 Hot Water Extractable Carbon (HWEC) 49
8.1 Introduction 49
8.2 Principle 50
8.3 Procedure 50
8.4 Calculation 50
Suggested Reading 52

9 Particulate Organic Matter-Carbon and Mineral Associated Organic Carbon 53
9.1 Introduction 53
9.2 Principle 54
9.3 Reagents 54
9.4 Equipment 54
9.5 Procedure 54
Suggested Reading 56

10 Determination of Water-Stable Aggregates Carbon 57
10.1 Introduction 57
10.2 Principle 58
10.3 Instruments 58
10.4 Procedure 58
10.5 Analysis of Sand-Free Carbon from Each Size Class 59
10.6 Calculation 61
Suggested Reading 61

11 Humic Acid (HA) and Fulvic Acid (FA) Carbon 63
11.1 Introduction 63
11.2 Principle 64
11.3 Reactions 64
11.4 Reagents 64
11.5 Procedure 64
11.6 Calculation 67
Suggested Reading 67

12 Determination of Soil Organic Carbon by Wet Digestion Method 69
12.1 Introduction 69
12.2 Principle 69
12.3 Equipment and Reagents 70
12.4 Procedure 70
12.5 Calculation 73
12.6 Critical Limit 73
Suggested Reading 73

13 Determination of Total Soil Organic Carbon (Dry Combustion Method) 75
13.1 Introduction 75
13.2 Principle 76
13.3 Equipment 76
13.4 Procedure 76
13.5 Note 78
Suggested Reading 78

14 Total Organic Carbon in Soil Extracts: Titration Method 79
14.1 Introduction 79
14.2 Principle 80
14.3 Reagents 80
14.4 Procedure 80
14.5 Calculation 81
Suggested Reading 81

15 Estimation of Total Carbon and Total Nitrogen in Soil Samples by Dumas Method 83
15.1 Introduction 83
15.2 Principle 84
15.3 Apparatus 84
15.4 Methodology 84

16 Soil Calcium Carbonate Equivalent by Volumetric Calcimeter Method 87
16.1 Introduction 87
16.2 Scope and Field of Application 88
16.3 Principle 88
16.4 Apparatus 88
16.5 Reagents and Standards 88
16.6 Sample Preparation 89
16.7 Procedure 89
16.7.1 Calibration of Calcimeter 89
16.7.2 Determination of the Total Carbonate in Soils 91
16.8 Sample Weight Estimation 91
16.9 Calculations 91
16.9.1 Calibration of Calcimeter 91
16.9.2 Determination of Carbonate in Soils 91
Suggested Reading 92

17 Assay of Soil Dehydrogenase Activity: TTC Method 93
17.1 Introduction 93
17.2 Principle 94
17.3 Apparatus 94
17.4 Reagents 94
17.5 Procedure 94
17.6 Standard Curve 95
17.7 Calculation 97
17.8 Example Calculation 97
Suggested Reading 97

18 Assay of Soil β-Glucosidase Activity 99
18.1 Introduction 99
18.2 Apparatus 100
18.3 Reagents 100
18.4 Procedure 100
18.4.1 Example Calculation 102
18.5 Standard Curve 102
18.6 Calculation 102
Suggested Reading 102

Series Editor and Editorial Board Members

About the Series Editor

Rajeev Kumar Varshney, FRS is a renowned agricultural scientist with over two decades of experience in genomics and molecular breeding. Currently, he holds multiple prestigious positions at Murdoch University in Australia. He has made a significant impact on global food security by creating genomic resources for major "orphan" tropical crops, using these resources to identify genetic loci and candidate genes for drought and pest tolerance in staple crops in sub-Saharan Africa and India, and spearheading international programs that deliver superior crop varieties to some of the world's most impoverished farmers. He is an accomplished author, with more than 600 publications to his name, and has been recognized as a Clarivate Highly Cited Researcher for ten consecutive years (2014–2023).

Food Futures Institute, Centre for Crop and Food Innovation, State Agricultural Biotechnology Centre, Agriculture and Food Security, Murdoch University, Perth, WA, Australia

Editorial Board Members

Trilochan Mohapatra Protection of Plant Varieties and Farmers' Rights Authority, Ministry of Agriculture and Farmers Welfare, Government of India, New Delhi, India

Soumya Swaminathan MS Swaminathan Research Foundation, Chennai, India

Wendy Umberger Australian Centre for International Agricultural Research (ACIAR), Canberra, Australia

Florence Wambugu Africa Harvest, Nairobi, Kenya

About the Authors

Dr. Santosh Kale Dr. Santosh Kale is an Associate Scientist at the International Crops Research Institute for the Semi-Arid Tropics (ICRISAT), with 7+ years of expertise in promoting waste-to-wealth practices that support regenerative agriculture techniques across a range of agro-ecologies, aiming to improve carbon sequestration at various production system levels. He has a doctorate in soil science and specializes in soil carbon dynamics, nutrient management, crop production, soil fertility, biochar, and soil carbon sequestration. He completed his Ph.D. from Mahatma Phule Krishi Vidyapeeth, Rahuri, India. He has authored over 25 research articles, book chapters, and popular articles

Dr. Rakesh S Dr. Rakesh S is an Associate Scientist at the International Crops Research Institute for the Semi-Arid Tropics (ICRISAT), working on the regenerative agriculture technologies across diverse agro-ecologies. He has 8+ years of experience in soil science research and has published more than 70 research articles in the reputed national and international journals. His research expertise includes conservation agriculture, carbon sequestration, carbon dynamics, soil fertility, plant nutrition, integrated nutrient management, land restoration, and climate change adaptation. He is acting as a Reviewer of Peer-Reviewed Journals of Elsevier, Springer, Frontiers, PLoS, MDPI, etc. He has an Editorial role for his Special Issue in Frontiers and also an Editorial Board Member in Resources, Environment & Sustainability-Elsevier.

Dr. Pushpajeet Choudhari Dr. Pushpajeet Choudhari is serving as a Soil Scientist at ICRISAT with over 8+ years of experience in managing the FAO-GLOSOLAN registered state-of-the-art laboratory. He chairs the Asian Soil Laboratory Network (FAO-SEALNET) and is known for harmonizing SOPs, ensuring quality and accuracy in soil and crop quality analysis, conforming to the international standards. His research work focuses on soil health monitoring, carbon sequestration, regenerative agriculture, and soil test–based nutrient management. He is at the forefront of leading bilateral/government-funded projects related to soil health monitoring, carbon sequestration, and sustainable soil health management. He has published research papers and book chapters in peer-reviewed journals and continues to mentor Ph.D. scholars, researchers in soil science.

Dr. Gajanan Sawargaonkar Dr. Gajanan Sawargaonkar is a Principal Scientist-System Agronomy with over 20+ years of experience in designing, planning, and scaling up regenerative agricultural practices across various cropping systems and agroecologies. He is leading the Regenerative Agri-Technologies sub-theme at ICRISAT. Dr. Sawargaonkar has led various research initiatives, in collaboration with a consortium of international and national organizations, focusing on cropping system design, management, and crop diversification/intensification. His efforts are directed towards building system resilience in dryland agro-ecologies, contributing to positive impacts on soil health, system productivity, profitability, and system resilience. Dr Sawargaonkar has published more than 45 research papers, 25 technical papers/research abstracts, and 30 extension articles.

Dr. Ramesh Singh Dr. Ramesh Singh currently holds the position of Principal Scientist & Global Theme leader – on Regenerative landscape under the Global Research Program on Resilient Farming Systems. He earned his Ph.D. from the GB Pant University of Agriculture & Technology, Pantnagar, India, and has 27 years of research and development experience in the areas of landscape resource conservation, following a land resource inventory and hydrology-based approach. He has high-level expertise in designing a range of cost-effective rainwater harvesting structure considering landscape terrain.

Dr. Mangi Lal Jat Dr. Mangi Lal Jat is a well-recognized Systems Agronomist with over 26 years of rich experience in systems science across the developing world. Currently, Dr. Jat serves as Secretary, Department of Agricultural Research and Education (DARE), and Director General of the Indian Council of Agricultural Research (ICAR), India, where he provides visionary leadership to India's national agricultural research and education system. Previously, as Global Research Program Director for the Resilient Farming system program at ICRISAT, he led multidisciplinary research teams across five clusters – climate adaptation and mitigation science, geo-spatial and big data sciences, digital agriculture, landscapes, soil health and water science, and the ICRISAT Development Center. His scientific, strategic, and partnership guidance has been pivotal in advancing resilience-focused agricultural innovations at global and regional scales.

Background

1

Abstract

This chapter provides a detailed overview of soil organic matter (SOM) and soil organic carbon (SOC), emphasizing their fundamental roles in soil fertility, nutrient cycling, and climate change mitigation. It discusses SOM as a vital reservoir of nutrients and energy for soil microorganisms, highlighting its impact on soil's physical, chemical, and biological health. The chapter also explores various carbon fractions, including particulate organic carbon (POC), dissolved organic carbon (DOC), microbial biomass carbon (MBC), mineral-associated organic carbon (MOC), and their significance in carbon farming and nutrient dynamics. Additionally, it highlights the importance of carbon sequestration in mitigating greenhouse gases and preserving the global carbon balance, referencing regenerative agriculture and the CSIFRA project by ICRISAT. The chapter concludes with laboratory safety guidelines for chemical handling, waste management, and equipment maintenance, ensuring safe and effective analytical practices. Collectively, these elements provide a solid methodological and conceptual foundation for measuring soil carbon, understanding its ecological roles, and promoting sustainable soil management in the context of climate change.

Keywords

Soil organic matter · Carbon fractions · Carbon farming · Regenerative agriculture · Nutrient cycling · Analytical chemistry · Soil sampling · Laboratory safety

S. Kale et al., *Soil Carbon Fractionation*, Sustainability Sciences in Asia and Africa,
https://doi.org/10.1007/978-981-95-4068-6_1

Soil organic matter (SOM) is crucial for maintaining soil fertility. SOM is a repository for essential plant nutrients and an energy source for soil microorganisms. It also functions as a sink for greenhouse gases (GHGs), controlling the exchange of gases between the ground surface and the atmosphere (Alengebawy et al. 2024). SOM storage in cultivating lands depends on climate, particularly annual precipitation, temperature, and cropping systems.

Soil organic carbon (SOC) plays a pivotal role in the functioning of agroecosystems which helps directly in the sustainable maintenance of soil health. It serves a sink for the long-term preservation of carbon (C) in the terrestrial ecosystem and contains two to three times greater C than the atmosphere C.

Soils in India contain comparatively little SOM, usually less than 0.5%, and it ranges from 0.1% to 1.0%, which has a significant effect on the physical status and fertility of the soil. SOM is typically lost when land is converted from a natural state to agriculture. In tropical soils, sustainable crop production depends on keeping SOC at a desired range of 0.5–1.0%. After a change in management, the SOM in temperate regions might take up to 50 years to reach a new equilibrium level. However, this time frame is somewhat shorter in tropical and semiarid regions like India. The SOM levels have significantly decreased in semiarid and sub-humid regions due to intensive farming and tillage practices. This results from inadequately protected soils being eroded by wind and water and enhanced microbial decomposition. A decrease in soil productivity is frequently linked to a decline in SOM levels. Because of the wide range of inputs it gets (including plant tissues, microbes, animals, and external organic waste) and the various phases of decomposition they go through, SOM is incredibly complicated (Chenu et al. 2024).

1.1 Carbon Fractions in Relation to Carbon Farming

Offsetting global greenhouse gas emissions (GHGs) has challenged humankind. The rise in atmospheric carbon dioxide (CO_2) levels contributes significantly to global warming, as a result, climate change issues impact agriculture at greater levels. Carbon farming offers a "win-win" opportunity for farmers and the environment. Enhancing SOC sequestration in soil systems through carbon farming practices is thus crucial. Landowners could produce offset credits for specific actions that raised carbon removal from the atmosphere or decreased greenhouse gas emissions; these credits could then be sold on both domestic and international markets.

Carbon farming helps increase soil organic carbon storage, which consists of various fractions such as particulate organic carbon (POC), dissolved organic carbon (DOC), light-fraction organic carbon (LFOC), mineral-associated organic carbon (MOC), microbial biomass carbon (MBC), short-term mineralizable carbon (Min-C), and permanganate oxidizable carbon (POXC). These fractions vary in formation, persistence, and function (Bongiorno et al. 2019; Singh et al. 2020). Understanding the process of SOC sequestration can be improved by studying changes in SOC fractions and the principles of SOC creation (Hu et al. 2023). Increasing SOC storage in agricultural soils can be achieved through conservation agriculture, which involves the combined use of organic and inorganic fertilizers,

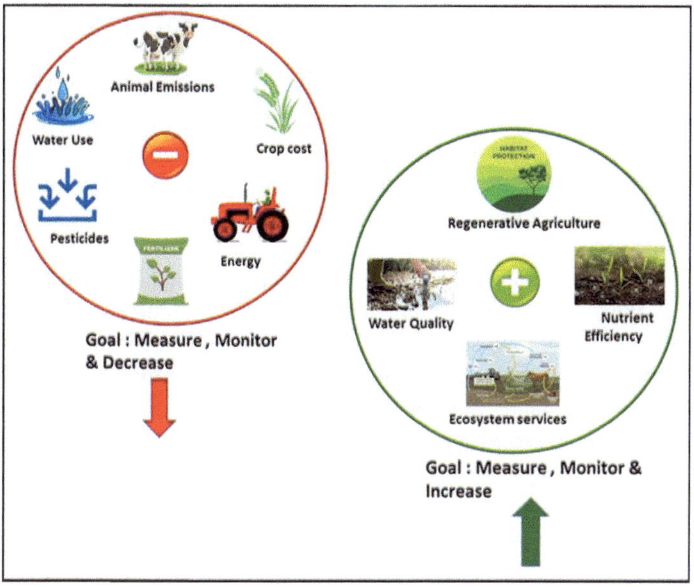

Fig. 1.1 Cropping carbon through regenerative agriculture. (Source: Project Leaflet published by ICRISAT)

mulching of crop residues or any carbon-rich biomass, planting cover crops, applying a balanced mix of macro- and micronutrients, using compost and biofertilizers, varying and diversifying crop systems, and promoting rhizobacteria (Upadhyay et al. 2019; Upadhyay and Edrisi 2021; Mahmud et al. 2021).

The International Crops Research Institute for the Semi-Arid Tropics (ICRISAT) has initiated a carbon farming project to develop and validate the Carbon Standards for Incentivizing Farmers for Regenerative Agricultural Practices (CSIFRA) (Fig. 1.1). The CSIFRA initiative aims to address climate change and advance regenerative agriculture while simultaneously leveraging social, economic, and environmental advantages. The key goals of a carbon farming project are to increase soil and plant carbon storage and absorb carbon dioxide (CO_2) from the environment through the adoption of regenerative agriculture practices such as (1) water management, (2) transplanting to direct seeding, (3) minimum soil disturbance, and (4) optimum use of fertilizer in combination with organics, crop residue, and biologicals. This supports smallholder farmers, as the institute is validating methods and instruments for measuring carbon footprints in agricultural systems and providing a data-driven monitoring, reporting, and verification (MRV) platform for supporting regenerative practices through carbon credit schemes.

1.2 Carbon Fractions in Relation to Nutrient Cycling

The SOC stock is made up of stable, resistant, and recalcitrant pools with varying residence times, as well as labile or actively cycling pools. The labile C pool represents the portion of SOC with quick turnover rates and is crucial for the soil food web as an energy source. It affects nutrient cycling and helps maintain soil quality and production (Chan et al. 2001). Microbial activity alters highly recalcitrant or passive pools very slowly (Sherrod et al. 2005). Active fractions (Fig. 1.2) such as MBC and MBN, POC and PON, and potentially mineralizable C and N can change the nutrient dynamic due to immobilization mineralization processes. This can better reflect variations in both soil quality and productivity, according to Shahid et al. (2017). These fractions also respond quickly to changes in management practices.

All organic compounds present in soils, regardless of their source or stage of decomposition, are referred to as SOM. It can include humus, DOC, POM, live organic matter (such as biomass from plants, animals, and microorganisms), and inert or highly carbonized organic matter (charcoal). The primary biological characteristics of organic matter are (i) its function as an energy source for faunal and microbial activities in the soil; (ii) its ability to stabilize enzyme activity; and (iii) its importance as a source of plant nutrition through mineralization. Because of their comparatively easy approach, SOM characteristics (microbial biomass C and N) could be utilized as markers of change because they are very sensitive to changes in

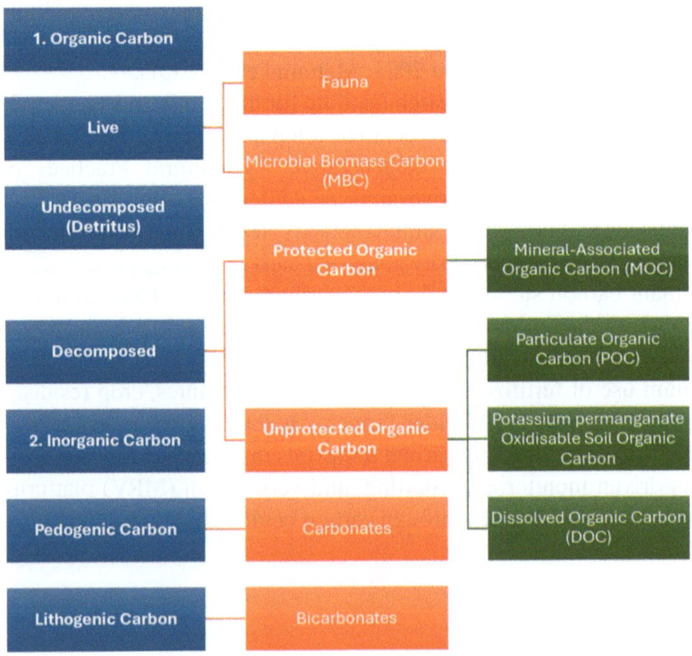

Fig. 1.2 Constituents of soil carbon pool

total soil organic matter (Eze et al. 2023). The functional SOM excludes organic materials that are more than 2 mm in size (Carter 2002). In contrast to soil organic/inorganic carbon, SOM is composed of nitrogen, phosphorus, sulfur, hydrogen, and carbon.

Soil carbon input, regulated by N availability and nutrients required to maintain atmospheric nitrogen (N_2) fixation, is the primary factor regulating C sequestration during plant development (Grover et al. 2015; Sawargaonkar et al. 2025; Khopade et al. 2025). The C:N ratio of agricultural residues and soil organic matter (SOM) controls soil processes. The typical C:N ratios of stable forms of soil organic carbon (SOC), humus, and soil microbial biomass (MB) are approximately 10–12, 10, and 7, respectively. By weight, SOM's C:N ratio is 12:1, its C:P ratio is 50:1, and its C:S ratio is 70:1. Therefore, to sequester 1 g of carbon, 83, 20, and 14 kg of nitrogen, phosphorus, and sulfur, respectively, are required (Singh et al. 2017). Additionally, Richardson et al. (2014) suggested that controlling inorganic nitrogen, phosphorus, and sulfur is an important aspect of techniques for creating soil carbon from high-carbon agricultural wastes.

1.3 Soil Carbon Sequestration and Climate Change Mitigation

Ecosystems are significantly impacted by the global climate change caused by escalating atmospheric carbon dioxide (CO_2) levels. Human activity and climate change have been significantly impacted by the rise in CO_2 levels brought on by human activity. The necessity of tackling this issue is highlighted by the rising temperatures, changing weather patterns, and related effects on ecosystems and human cultures. Reduced agricultural yields, increased food insecurity, and broader economic disparities are the outcomes of climate change's significant influence on human activity and detrimental effects on agriculture.

Soil is a huge global carbon sink that stores twice as much carbon as the atmosphere and nearly three times as much as plants (Ramesh et al. 2019). This process of storage of atmosphere CO_2 in the soil system as C is known as C sequestration. The amount of carbon that a soil system can store is determined by how well net gains and net losses are balanced. Before the dramatic increase in C emissions that occurred during the industrial revolution, the global carbon cycle, also known as the "C flux," was kept practically balanced between the consumption of CO_2 (sinks) and its release back into the atmosphere (sources). Even a small alteration in soil carbon sequestration will significantly affect the global carbon cycle and climate change. The SOC is made up of various pools and fractions, including TOC, POC, DOC, MBC, POXC, MOC, and others, which are all integral to carbon sequestration. However, they differ in disintegration rate and stability.

Given the 1.5 °C target set by the 2015 Paris Agreement, it has become essential to remove atmospheric carbon dioxide. Integrated Assessment Models (IAMs) suggest that between 190 and 1190 $GtCO_2$ of cumulative removal may be required by the end of the century, depending on the extent and pace of climate change mitigation actions (Huppmann Ranjan et al. 2017). It is crucial to reduce GHG emissions quickly and potentially draw more carbon from the atmosphere to limit global warming to less than 2 °C. Increasing SOC stock is significant because it affects crop yield and contributes to achieving climate change and sustainable development goals (Srinivasarao et al. 2021).

1.4 Basic Principles in Analytical Chemistry

A comprehensive grasp of foundational concepts is essential in analytical chemistry. This section briefly outlines several key principles and definitions that are significant for quantitative analysis.

Mass and Weight
"Mass" refers to the amount of matter in a substance and is a fundamental property measurable with a balance. In contrast, "weight" is the force exerted by gravity on that substance. Although "weight" is technically a measure of force, it is often used interchangeably with "mass" in everyday language.

Atom and Atomic Weight
An atom is the smallest unit of an element that retains its chemical properties and can engage in chemical reactions. Atomic weight is the weighted average of the masses of the isotopes of an element, based on a scale where the mass of a carbon-12 atom is exactly 12 atomic mass units (amu). When expressed in grams, the atomic weight corresponds to one mole of the element, often referred to as a gram atom. For example, 16 g of oxygen represents 1 g atom of oxygen.

Avogadro's Number
Avogadro's number is the number of constituent particles (atoms, molecules, ions, etc.) in one mole of a substance, and its value is 6.022×10^{23}. This constant also represents the number of atoms in 1 g atom of any element.

Molar Volume
At standard temperature and pressure (0 °C and 1 atmosphere), one mole of any gaseous substance occupies a volume of 22.4 L. This is known as the molar volume.

Molecule
The molecule is the smallest unit of a chemical compound that might exist separately and maintain the chemical characteristics of the given compound. It comprises two or more bound atoms arranged specifically.

Molecular Weight
Sometimes referred to as molecular mass, molecular weight is the total of all the atoms' atomic weights in a specific molecule. Usually expressed in atomic mass units, it is rather helpful in determining the content of a specific sample.

Standard Solution

A standard solution is a solution with a precisely known concentration, typically expressed in terms of normality. It is prepared by dissolving a known weight of a reagent in a specific volume of solvent. Standard solutions can be classified as primary or secondary standards. Primary standards are highly pure substances that are stable and do not change composition during storage or use, making them suitable for preparing solutions with known concentrations.

For acid solution: Succinic acid, Oxalic acid
For alkali solution: $Na_2B_4O_7$ $10H_2O$, Na_2CO_3 (anhydrous)
For redox system: Sodium oxalate, Potassium dichromate, etc.

Comparative Weight: The equivalent weight of a material is the amount of that substance that reacts with or displaces 8.00 parts by weight of oxygen, 1.008 parts by weight of hydrogen, or 35.5 parts by weight of chlorine. The gram equivalent weight is the expression of this amount in grams.

Concentration of Solutions: A solution's concentration indicates how much solute has been dissolved in a given volume. This can be expressed in various ways.

Normality: One gram equivalent of the solute dissolved in 1 L of solvent creates a normal solution. Normality (N) is the concentration of the solute expressed in gram equivalents per liter of solution. For instance:

- *1 N solution*: 1 g equivalent per liter
- *N (decinormal)*: 0.1 g equivalent per liter
- *N (centinormal)*: 0.01 g equivalent per liter
- *N (millinormal)*: 0.001 g equivalent per liter

Molal Solution and Molality

Represented as mol/kg, a molal solution is one mole of solute dissolved in 1000 g of solvent instead of in the solution volume.

Molar Solution and Molarity

Represented as M, a molar solution has 1 g molecular weight of solute dissolved in 1 L of fluid. Molarity, or molarity (M), is defined as 1 L of solution's moles of solute present.

Composition by Percentage Weight

This approach shows concentration by mass of solute per 100 g of solution. A 10% KCl solution, for instance, produces 100 g of the resultant solution by mixing 10 g of potassium chloride in 90 g of water.

Table 1.1 Molecular and equivalent weights chemicals

S. No.	Compound	Formula	Mol. wt. (g)	Eq. wt. (g)
1	Ferrous sulfate	$FeSO_4.7H_2O$	278.0	139.0
2	Potassium hydroxide	KOH	56.1	56.1
3	Ammonium nitrate	NH_4NO_3	80.0	80.0
4	Sodium chloride	$NaCl$	58.5	58.5
5	Sodium sulfate	Na_2SO_4	142.0	71.0
6	Sodium carbonate	Na_2CO_3	106.0	53.0
7	Potassium sulfate	K_2SO_4	174.3	87.1
8	Sodium bicarbonate	$NaHCO_3$	84.0	84.0
9	Sodium acetate (anhydrous)	CH_3COONa	82.0	82.0
10	Sodium oxalate	Na_2C2O_4	134.0	67.0
11	Potassium chloride	KCl	74.6	74.6
12	Oxalic acid	$C_2H_2O_4.2H_2O$	126.0	63.0
13	Barium acetate	$(CH_3COO)_2$ Ba	255.4	127.7
14	Ammonium fluoride	NH_4F	37.0	37.0
15	Potassium hydrogen phthalate	$COOHC_6H_4COOK$	204.2	204.2
16	Silver nitrate	$AgNO_3$	169.9	169.9
17	Calcium hydroxide	$Ca(OH)_2$	74.0	37.0
18	Sodium nitrate	$NaNO_3$	85.0	85.0
19	Potassium nitrate	KNO_3	101.1	101.1
20	Sodium hydroxide	$NaOH$	40.0	40.0
21	Potassium permanganate	$KMnO_4$	158.0	31.6
22	Potassium dichromate	$K_2Cr_2O_7$	294.2	49.0
23	Boric acid	H_3BO_3	61.8	20.6
24	Ferrous ammonium sulfate	$(NH_4)_2SO_4.FeSO_4.6H_2O$	392.1	392.1
25	Calcium acetate	$(CH_3COO)_2$ Ca	158.0	79.0
26	Sodium thiosulfate	$Na_2S_2O_3.5H_2O$	248.2	248.2
27	Ammonium chloride	NH_4Cl	53.5	53.5
28	Magnesium nitrate	$Mg(NO_3)_2.6H_2O$	256.4	128.2
29	Ammonium acetate	CH_3COONH_4	77.1	77.1
30	Magnesium chloride	$MgCl_2.6H_2O$	203.3	101.7
31	Barium chloride	$BaCl_2.2H_2O$	244.3	122.1
32	Calcium nitrate	$Ca(NO_3)_2$	164.0	82.0
33	Calcium chloride (dehydrate)	$CaCl_2.2H_2O$	147.0	73.5
34	Calcium carbonate	$CaCO_3$	100.1	50.1
35	Calcium sulfate	$CaSO_2.2H_2O$	172.2	86.1

Formal Solution

The formula weight of the solute per liter of solution defines a formal solution. It is designated F. Usually equal to the molecular weight, the formula weight may occasionally be a multiple of the actual molecular weight, particularly in relation to chemical reactions (Tables 1.1 and 1.2; Fig. 1.3).

Table 1.2 Percent concentration, specific gravity, normality, and amount needed for making 1 N solution of some commonly used liquid reagents

Reagent	Conc. (%w/w)	Sp. Gr. (g mL^{-1})	Normality (approx.)	Quantity (mL) required for 1 L of 1 N solution
Sulfuric acid	96	1.84	36	28
Phosphoric acid	85	1.7	44	23
Ammonium hydroxide	28.3	0.91	15	67
Acetic acid glacial	99	1.06	17.5	58
Nitric acid	71	1.42	15.6	62
Hydrochloric acid	39	1.19	11.8	78

Fig. 1.3 Potassium dichromate solution for organic carbon estimation

Percentage Composition of Volume

The concentration is articulated as the volumes of solute and solvent. A 20% ethanol solution combines 20 mL of ethanol with 80 mL of water.

Strength or Percentage Strength

The strength of a solution is defined as the mass of solute, in grams, dissolved in 1 L of solution. It is typically denoted as g/L. Percentage strength denotes the quantity of solute, in grams, per 100 mL of solution.

Equivalence Point or Endpoint

The point at which a titration reaches its conclusion is known as the equivalence point or endpoint. At this stage, the titrant has wholly reacted with a known amount of the analyte, and the reaction is quantitatively complete at this specific volume.

Parts per Million (ppm)

Concentration is often measured as the amount of solute in grams per 1 million mL of solution or milligrams per liter. For instance, a solution containing 10 mg of solute per liter, or 10 μg of solute per milliliter, has a concentration of 10 parts per million (ppm). Another standard unit of concentration is milliequivalents per liter (meq/L), which represents the quantity of a substance as one-thousandth of a gram equivalent per liter of solution.

Indicator

An indicator is a substance that signals the completion of a reaction. It is an auxiliary reagent that helps visually determine when a titration has reached its endpoint. Three types of indicators are commonly used in volumetric analysis.

Acid-Base Indicators

These include Methyl orange, Phenolphthalein, Methyl red, Bromothymol blue, phenol red, and Thymolphthalein. They are known as hydrogen ion or acid-base indicators, organic compounds that change color within specific pH ranges when added to a solution. Each indicator has a defined pH range for its color change, which occurs gradually within a narrow interval, known as the color change interval. Selecting an appropriate indicator in acid-base titrations is crucial, as it should show a distinct color change close to the pH of the equivalence point (Table 1.3).

Indicators Used in Oxidation-Reduction Reactions (Redox Indicators)

A chemical that shows different colors in its reduced and oxidized states is called an oxidation-reduction (Redox) indicator. This group includes certain organic dyes. The chosen indicator should have a standard potential (Eo) between the standard potentials of the titrated oxidation-reduction systems to clearly display a color shift at the endpoint. Examples of these indicators include diphenylamine,

Table 1.3 pH ranges of various indicators and their color changes in acid and alkaline medium

Indicators	pH	Acid medium	Alkaline medium
Phenol red	6.4–8.2	Yellow	Red
Methyl orange	3.1–4.5	Red	Yellow
Thymolphthalein	9.3–10.5	Colorless	Blue
Litmus	5.0–8.0	Red	Blue
Methyl red	4.2–6.3	Red	Yellow
Phenolphthalein	8.0–10.0	Colorless	Pink
Bromothymol blue	6.0–7.6	Yellow	Blue

Table 1.4 Some redox indicators

Indicator	Color		E^{o} at $H^{+} = 1$ (V)
	In (Ox)	In (Red)	
Neutral red	Red	Colorless	+0.24
Diphynlamineazosulfonic acid	Red-Violet	Colorless	+0.85
Erioglucin A	Red	Green	+1.00
Methylene blue	Blue	Colorless	+0.53
Indigo tetrasulfonate	Blue	Colorless	+0.36
Diphenylamine	Blue-Violet	Colorless	+0.76
Ferroin (Fe^{++}orthrophenanthroline)	Pale blue	Red	+1.14
Phenylanthranilic acid	Red-Violet	Colorless	+1.08
Nitroferroin	Pale blue	Red	+1.25
Diphenylamine-2,2′-dicarboxylic acid	Blue-Violet	Colorless	+1.26

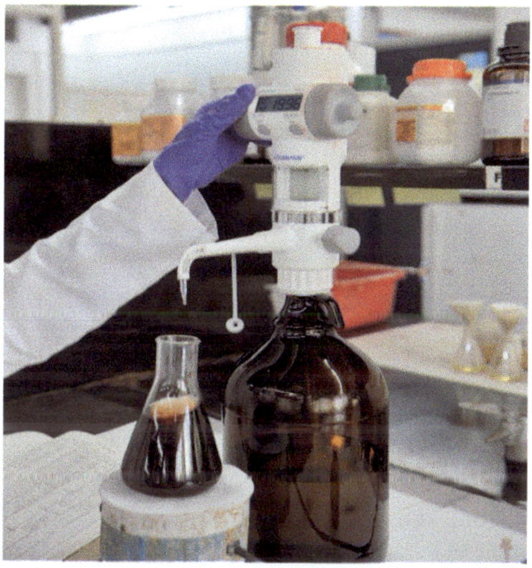

Fig. 1.4 Advance titrator equipment for precise determination of organic carbon through wet oxidation. (The Charles Renard Analytical Laboratory (CRAL), ICRISAT)

diphenylbenzidine, ortho-phenanthroline, methylene blue, starch, Eriochrome Black T, and murexide (Table 1.4; Fig. 1.4).

Self-Indicator

A chemical is considered a self- or auto-indicator when it causes a visual color change. Each time potassium permanganate ($KMnO_4$) is titrated in diluted sulfuric acid with oxalic acid or ferrous sulfate, it acts as a self-indicator. The formation of colorless products such as $MnSO_4$, K_2SO_4, and H_2O leads to decolorization, and a distinct pink color indicates the endpoint.

Ions and Ionization

When electrolytes dissolve in water, they dissociate into charged entities called ions. The process known as ionization generates positively charged ions (cations) and negatively charged ions (anions). This dissociation is essential for electrical conduction in solutions and is integral to other chemical processes.

Oxidation-Reduction (Redox) Reactions

Oxidation denotes the process by which a substance relinquishes electrons, elevating its oxidation state. Conversely, reduction entails the acquisition of electrons, diminishing its oxidation state. In a redox reaction, these two processes invariably co-occur, as the electrons relinquished by one species are acquired by another. Redox reactions are essential in analytical chemistry for ascertaining the composition of various substances, as they reveal the existence and concentration of specific elements or compounds in a solution.

Example: Ferrous is oxidized to ferric by losing one electron, and simultaneously, $KMnO_4$ is reduced by taking five electrons.

Oxidation

$$Fe^{++} \underset{-e^-}{\overset{-2e^-}{\rightarrow}} Fe^{+++} \quad (\text{Loss of electrons})$$

$$\mathrm{C_2O_{4^-}^-} \rightarrow 2CO_2$$

Reduction

$$Cl_2 \underset{+2e^-}{\overset{+5e^-}{\rightarrow}} 2Cl^- \quad (\text{gainofelectrons})$$

$$MnO_4^- + 8\mathrm{H}^+ \rightarrow Mn^{++} + 4\mathrm{H_2O}$$

Oxidation Number

The oxidation number is the number of electrons lost or gained in a given reaction by one substance molecule.

Common-Ion Effect

The common-ion effect refers to the phenomenon where the dissociation of a weak acid or weak base is reduced upon adding a compound containing an already present ion. For example, the dissociation of ammonium hydroxide (NH_4OH) is suppressed when ammonium chloride (NH_4Cl) is added, as both contain the common ion NH_4^+. This effect is particularly useful in qualitative analysis for the separation and identification of certain metal ions, such as Fe^{3+}, Al^{3+}, and Cr^{3+}, in group III of the periodic table.

1.5 Soil Sampling and Processing

Why Soil Testing?

Assessing soil fertility involves estimating the available nutrients in the soil, which are directly accessible for crop plants to absorb. This assessment helps determine the optimal ratio for applying fertilizers. It also evaluates the soil's nutrient status and identifies any deficiencies. Additionally, it considers the environmental quality and potential community hazards.

Objectives

(i) To estimate the level of nutrients (Low, Medium, High) in order to assess soil fertility and productivity.
(ii) To identify any specific soil issues, such as acidity, alkalinity, and sodicity. Following that, suggestions for their rectification (such as the need for gypsum or lime) will be made.
(iii) Estimating the likelihood that crops will respond to fertilizers to the greatest extent possible.

Sampling Equipment

Spade, screw auger (depth-wise) or post-hole auger (for moist soil), Kudali (pick-axe), khurpi, trowel, plastic container, steel spatula, clean plastic bucket. Cloth bag, information sheet, scale for measuring pit depth, and marking pencil or ink.

How to Collect Soil Samples

- Remove or scrape the upper portion or surface of the soil and remove the crop residues and litter on the soil surface.
- Take kudali and spade with a spatula and dig out near about 15–20 cm deep soil with kudali or pick axe. The shape should be V-shaped. Remove and reject the first dug surface soil. Then, the soil is collected in plastic paper or containers from selected spots in the field.
- Mix all the soil from different spots in the plastic paper or container.
- Avoid soil sampling where fertilizers and manure have been applied recently. After mixing the soil sample, divide it into four parts. Then, remove two parts from the quarter and take the remaining two parts of the soil from the plastic paper or container.
- Minimum: We must take ½ kg or 1 kg soil sample for analysis. Write the information which is given in the sheet (name of the farmer, village, tehsil, district, survey number, earlier crop taken, soil type, and next crop)

Soil Sampling for Suitability of Horticultural Crops

Decide representative one spot of soil sampling. Depth-wise soil sampling should be done as 0–30 cm (first layer), 30–60 cm (second layer), and 60–90 cm depth (third layer). Collect soil samples layer-wise and send them to the laboratory after drying and labeling.

Drying

Spread the soil sample evenly on a clean polythene sheet and allow it to air-dry for 4–5 days in a well-ventilated area, such as a room or shed. Ensure the sample is kept free from contamination during the drying process.

Grinding

Grind the air-dried soil using a mortar and pestle to break down any clumps and to homogenize the sample. Proper grinding ensures that the soil particles are uniform in size, which is crucial for accurate analysis.

Sieving

Pass the ground soil through a 2 mm sieve to obtain a fine and consistent sample. Avoid using brass, copper, or galvanized iron sieves to prevent contamination with micronutrients, which could interfere with the analytical results.

Factors Affecting Soil Carbon Dynamics

Soil organic carbon (SOC) dynamics are governed by complex interactions among climate, soil properties, land use, and management practices. Understanding these factors is essential for improving soil fertility, predicting carbon sequestration potential, and formulating sustainable land management strategies.

Climatic Factors: Temperature and precipitation strongly influence the rate of organic matter decomposition and stabilization. High temperatures accelerate microbial activity, enhancing SOC mineralization, while moisture availability determines both the input of plant residues and microbial processes. In semi-arid regions, limited rainfall and high evaporation often result in slower carbon accumulation compared to humid zones (Meena et al. 2025; Rakesh et al. 2023).

Soil Properties: The texture, mineralogy, and pH of soil are key determinants of SOC stabilization. Clay-rich soils tend to protect organic matter within microaggregates, reducing its decomposition rate. In contrast, sandy soils with lower surface area exhibit rapid turnover and limited storage capacity. The presence of calcium and iron oxides further promotes carbon stabilization through organo-mineral complexes (Sawargaonkar et al. 2024).

Land Use and Vegetation: Changes in land use such as conversion of forest to cropland lead to rapid SOC depletion due to reduced litter input and increased disturbance. Native vegetation with deep root systems contributes more stable carbon inputs than annual crops. Restoring degraded lands through afforestation or perennial grasses significantly enhances long-term carbon sequestration (Kale et al. 2025).

Agricultural and Management Practices: Tillage intensity, residue management, crop rotation, and organic amendments directly influence SOC levels. Conservation agriculture, minimum tillage, cover cropping, and biochar application improve soil structure and carbon retention by promoting microbial diversity and aggregate stability (Jat et al. 2021; Bongiorno et al. 2019). Balanced nutrient management and the integration of organic and inorganic fertilizers further enhance carbon input efficiency (Kumar et al. 2024).

Topography and Erosion: Slope position affects soil erosion and redistribution of carbon. Soils on upper slopes often lose fine particles and SOC through runoff, whereas lower positions tend to accumulate sediments rich in organic matter. Implementing soil- and water-conservation measures helps minimize such losses (Pasumarthi et al. 2024).

1.6 Laboratory Safety and Handling Practices

Ensuring laboratory safety is an integral part of the analysis. The procedures described in this manual involve the use of strong acids, oxidizing agents, and organic solvents that require careful handling to prevent injury, contamination, or damage to equipment. Adherence to the following guidelines will help maintain a safe and sustainable laboratory environment.

General Safety Measures

All laboratory personnel must receive appropriate safety training before performing analytical procedures. Personal protective equipment (PPE) including a laboratory coat, safety goggles, gloves, and closed footwear must be worn at all times. Long hair should be tied back and loose clothing or jewelry should be avoided. Eating, drinking, or storing food inside the laboratory is strictly prohibited.

Chemical Handling

Strong acids (e.g., H_2SO_4, HCl, HNO_3) and oxidizing agents (e.g., $K_2Cr_2O_7$, $KMnO_4$) should be handled inside a fume hood. Always add acid to water, never the reverse, to prevent violent reactions. Use only clean and properly labeled glassware. Prepare standards and reagents with distilled or deionized water to minimize contamination. In case of chemical spills, neutralize acid residues with sodium bicarbonate and clean immediately while wearing appropriate PPE.

Waste Management

Chemical waste should be segregated into acid, base, and organic waste containers. Never dispose of reagents or heavy-metal residues directly into the drainage system. Potassium dichromate and permanganate wastes must be collected separately for reduction and safe disposal through an authorized hazardous-waste management facility. Encourage the reuse of glassware and containers after proper cleaning to minimize waste generation.

Fire and Electrical Safety

Keep flammable solvents away from heat sources and open flames. Ensure that fire extinguishers, eyewash stations, and first-aid kits are easily accessible. Electrical instruments should be checked periodically for earthing, insulation, and functional integrity. Switch off analytical instruments and exhaust systems when not in use.

Suggested Reading

Alengebawy A, Deka TJ, Pradhan SP, Elshazly A, ElSayed EE, Elgarhy AH (2024) Soil as a natural source of greenhouse gas emissions Advances and Technology Development in Greenhouse Gases: Emission, Capture, and Conversion (pp. 227–263)

Benedetti F, Caon L (2021) Global soil laboratory assessment 2020 – laboratories' capacities and needs. FAO, Rome

Bhattacharyya T, Ray SK, Gangopadhaya SK, Wani SP (2025) Black soils and their carbon sequestration potential Indian examples. Indian J Fert 21(8):712–733

Bongiorno G, Bünemann EK, Oguejiofor CU, Meier J, Gort G, Comans R et al (2019) Sensitivity of labile carbon fractions to tillage and organic matter management and their potential as comprehensive soil quality indicators across pedoclimatic conditions in Europe. Ecol Indic 99:38–50

Carter MR (2002) Soil quality for sustainable land management: organic matter and aggregation interactions that maintain soil functions. Agron J 94(1):38–47

Chan KY, Bowman A, Oates A (2001) Oxidizable organic carbon fractions and soil quality changes in an oxic paleustaff under different pastures leys. Soil Sci 166:61–67

Chenu C, Rumpel C, Védère C, Barré P (2024) Methods for studying soil organic matter: nature, dynamics, spatial accessibility, and interactions with minerals. In Soil microbiology, ecology and biochemistry (pp. 369–406). Elsevier.

Estefan G, Sommer R, Ryan J (2013) Methods of soil, plant, and water analysis. A manual for the West Asia and North Africa region, vol 3, issue 2. International Center for Agricultural Research in the Dry Areas (ICARDA), Beirut, pp 65–119

Eze S, Magilton M, Magnone D, Varga S, Gould I, Mercer TG, Goddard MR (2023) Meta-analysis of global soil data identifies robust indicators for short-term changes in soil organic carbon stock following land use change. Sci Total Environ 860:160484

Grover M, Maheswari M, Desai S, Gopinath K, Venkateswarlu B (2015) Elevated CO2: plant associated microorganisms and carbon sequestration. Appl Soil Ecol 95:73–85

Hu Q, Thomas BW, Powlson D, Hu Y, Zhang Y, Jun X et al (2023) Soil organic carbon fractions in response to soil, environmental and agronomic factors under cover cropping systems: a global meta-analysis. Agric Ecosyst Environ 355:108591

Huppmann Ranjan P, Patle GT, Prem M, Solanke KR (2017) Organic mulching-a water saving technique to increase the production of fruits and vegetables. Curr Agric Res J 5(3):371–380

Jat HS, Datta A, Choudhary M, Sharma PC, Dixit B, Jat ML (2021) Soil enzymes activity: Effect of climate smart agriculture on rhizosphere and bulk soil under cereal based systems of northwest India. Eur J Soil Biol 103:103292

Jones J (2018) Soil analysis handbook of reference methods. CRC Press
Kale S, Bhakare B, Kausadikar H, Thakare R, Durgude S, Rakesh S (2025) Impact of land use systems on soil chemical properties in semi-arid tropics of western Maharashtra, India. Int J Environ Clim Chang 15(6):98–109. https://doi.org/10.9734/ijecc/2025/v15i64876
Khopade R, Sawargaonkar G, Kale S, Davala M, Rakesh S, Das SS et al (2025) Sustainable intensification in coconut for building system resilience and nutritional security of smallholders in Eastern India. npj Sustainable Agric 3(1):42
Kumar NV, Sawargaonkar G, Rani CS, Rajesh Pasumarthi R, Kale S, Prakash TR, Triveni S, Singh A, Davala MS, Khopade R, Karthik R, Venkatesh B, Chandra MS (2024) Harnessing the potential of pigeonpea and maize feedstock biochar for carbon sequestration, energy generation, and environmental sustainability. Bioresour Bioprocess 11(1):5. https://doi.org/10.1186/s40643-023-00719-3
Lal R, Negassa W, Lorenz K (2015) Carbon sequestration in soil. Curr Opin Environ Sustain 15:79–86
Mahmud AA, Upadhyay SK, Srivastava AK, Bhojiya AA (2021) Biofertilizers: a Nexus between soil fertility and crop productivity under abiotic stress. Curr Res Environ Sustain 3:100063
Meena S, Manjaiah KM, Sharma VK, Purakayastha TJ, Das S, Bana RS et al (2025) Exploring soil organic carbon fractions, stocks, and carbon management index across land uses in subtropical ecosystems of Tripura, India. Front Sustain Food Syst 9:1604101
Okalebo, J. R., Gathua, K. W., & Woomer, P. L. (2002). Laboratory methods of soil and plant analysis: a working manual, 2nd edn, vol 21. Sacred Africa, Nairobi, pp 25–26
Pasumarthi R, Sawargaonkar G, Kale S, Kumar NV, Choudhari PL, Singh R, Jat ML (2024) Innovative bio-pyrolytic method for efficient biochar production from maize and pigeonpea stalks and their characterization. J Clean Prod 448:141573. https://doi.org/10.1016/j.jclepro.2024.141573
Rakesh S, Sinha AK, Sarkar D, Roy D, Bodiga D, Sahoo S, Rakshit A (2023) Active and passive carbon fractions in contrasting cropping systems, tillage practices, and soil types. Land 12(2):365. https://doi.org/10.3390/land12020365
Ramesh T, Bolan NS, Kirkham MB, Wijesekara H, Manjaiah KM, Srinivasarao C, Sandeep S, Rinklebe J, Ok YS, Choudhury BU, Want H, Tang C, Song Z, Freeman OW II (2019) Soil organic carbon dynamics: impact of land use changes and management practices: a review. Adv Agron 156:1–125
Richardson EA, Kirkby AC, Banerjee S, Kirkegaard J (2014) The inorganic nutrient cost of building soil carbon. Carbon Manag 5:265–268
Sawargaonkar G, Pasumarthi R, Kale S, Choudhari P, Rakesh S, Mutnuri S, Jat ML (2024) Valorization of peanut shells through biochar production using slow and fast pyrolysis and its detailed physicochemical characterization. Front Sustain 5:1417207. https://doi.org/10.3389/frsus.2024.1417207
Sawargaonkar GL, Kamdi P, Kale S, Shinde S, Das SS, Kausadikar H, Singh R, Jat ML (2025) Leveraging genotypex management synergies to enhance pigeonpea productivity, profitability, and sustainability in semi-arid tropics. Front Agron 7:1590421
Shahid M, Nayak AK, Puree C, Tripathi R, Lal B, Gautam P, Bhattacharyya P, Mohanty S, Kumar A, Panda BB et al (2017) Carbon and nitrogen fractions and stocks under 41 years of chemical and organic fertilization in a sub-humid tropical rice soil. Soil Tillage Res 170:136–146
Sherrod LA, Peterson GA, Westfall DG, Ahuja LR (2005) Soil organic carbon pools after 12 years in no-till dryland agroecosystems. Soil Sci Soc Am J 67:1533–1543
Srinivasarao C, Kundu S, Yashavanth BS, Rakesh S, Akbari KN, Sutaria GS, et al (2021) Influence of 16 years of fertilization and manuring on carbon sequestration and agronomic productivity of groundnut in vertisol of semi-arid tropics of Western India. Carbon Manag 12(1):13–24
Singh G, Sangwan S, Narwal E, Annapurna K (2017) Microbial proteins and soil carbon sequestration. In: Rakshit A (ed) Adaptive soil management: from theory to practices. Springer, Singapore, pp 131–139

Singh S, Nouri A, Singh S, Anapalli S, Lee J, Arelli P, Jagadamma S (2020) Soil organic carbon and aggregation in response to thirty-nine years of tillage management in the southeastern US. Soil Tillage Res 197:104523

Upadhyay SK, Edrisi SA (2021) Developing sustainable measures to restore fly ash contaminated lands: current challenges and future prospects. Land Degrad Dev 32:4817–4831

Upadhyay SK, Saxena AK, Singh JS, Singh DP (2019) Impact of Native ST-PGPR (Bacillus pumilus; EU927414) on PGP traits, antioxidants activities, wheat plant growth and yield under salinity. Clim Chang Environ Sustain 7:157

Permanganate Oxidizable Carbon (POXC) 2

Abstract

The potassium permanganate oxidizable carbon (POXC) method is extensively utilized and offers a swift means to assess the labile percentage of soil organic carbon (SOC). This percentage exhibits heightened sensitivity to managerial alterations and environmental factors. POXC denotes the active, readily decomposable organic matter in the soil, providing insights into soil quality and carbon cycles. This method provides a dependable and economical means to track variations in soil organic matter, making it an essential instrument for evaluating sustainable agriculture practices and soil health management programs (Weil et al. 2003). This procedure outlines a method for measuring POXC in soil samples, based on the "Active Carbon" method by Weil et al. (2003), with adaptations from the Glover Lab at the Land Institute and the Barbercheck Lab at Pennsylvania State University.

Keywords

Permanganate oxidizable carbon (POXC) · Labile soil organic carbon · Active carbon · Potassium permanganate method · Soil carbon fractionation · Sustainable agriculture · Soil health assessment

2.1 Introduction

The potassium permanganate oxidizable carbon (POXC) method is extensively utilized and offers a swift means to assess the labile percentage of soil organic carbon (SOC). This percentage exhibits heightened sensitivity to managerial alterations and environmental factors. POXC denotes the active, readily decomposable organic matter in the soil, providing insights into soil quality and carbon cycles. This method provides a dependable and economical means to track variations in soil organic

S. Kale et al., *Soil Carbon Fractionation*, Sustainability Sciences in Asia and Africa,
https://doi.org/10.1007/978-981-95-4068-6_2

matter, making it an essential instrument for evaluating sustainable agriculture practices and soil health management programs (Weil et al. 2003). This procedure outlines a method for measuring permanganate oxidizable carbon (POXC) in soil samples, based on the "Active Carbon" method by Weil et al. (2003), with adaptations from the Glover Lab at the Land Institute and the Barbercheck Lab at Pennsylvania State University.

2.2 Principle

The POXC method is based on the oxidation of the active (labile) portion of soil organic carbon by potassium permanganate ($KMnO_4$). The soil's labile organic carbon, which is more readily broken down and contributes significantly to the cycling of nutrients, combines with $KMnO_4$ to change its hue from purple (Mn^{7+}) to colorless or light brown (Mn^{2+}). The change in absorbance or color intensity of the $KMnO_4$ solution, which is determined spectrophotometrically at 550 nm, is proportional to the amount of oxidizable organic carbon. $KMnO_4$ reacts with the more easily oxidizable forms of soil C; Mn(VII) being reduced to Mn(II) correspondingly lowers absorbance at 550 nm. The chemical reaction can be characterized as follows:

$$5C_2O_4{}^{2-}(aq) + 2MnO_4 - (aq) + 16H^+(aq) \rightarrow 10CO_2(g) + 2Mn^{2+}(aq) + 8H_2O(l)$$

2.3 Instrumentation and Materials

- Spectrophotometer
- Weighing balance
- pH meter
- Horizontal shaker
- Magnetic stir plate
- Polypropylene centrifuge tubes
- Potassium permanganate ($KMnO_4$; FW = 158.03 g/mol)
- Calcium chloride, dihydrate ($CaCl_2 \cdot 2H_2O$; FW = 147.01 g/mol)

I. *Reagent Preparation: $KMnO_4$ Stock Solution 0.2 M (makes 1 L)*:

1. Weigh 147 g of $CaCl_2$ and add it to a 1000 mL beaker with 900 mL deionized water. Stir until dissolved; no heating is required. Transfer to a volumetric flask or graduated cylinder and bring to 1 L with deionized water.
2. Weigh 31.6 g of $KMnO_4$ and add it to a 1000 mL beaker containing 900 mL of the $CaCl_2$ solution. Stir gently with heat until dissolved completely. You may need to decant some solution to check for undissolved $KMnO_4$.
3. Measure the pH with a calibrated meter while stirring. Adjust to 7.2 using 0.1 N NaOH, adding one drop at a time, and allowing the reading to stabilize. Once adjusted, transfer to a 1000 mL flask and fill to volume with $CaCl_2$ solution.

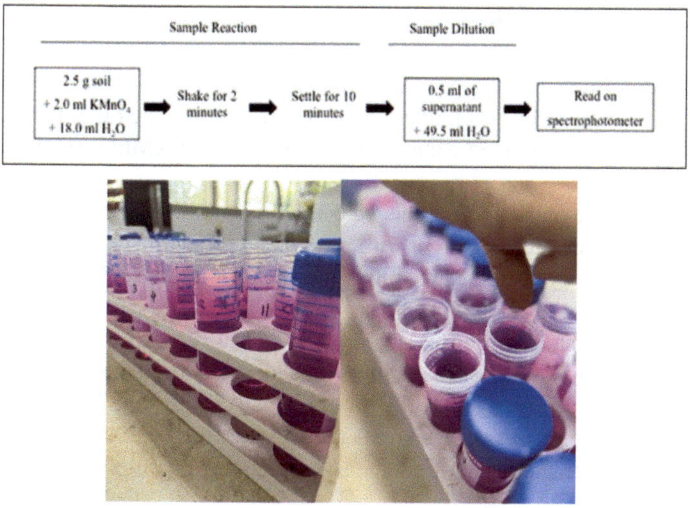

Fig. 2.1 Potassium permanganate oxidizable soil carbon

4. Adjust the $KMnO_4$ solution volume based on the number of samples; each soil sample requires 2.0 mL of 0.2 M $KMnO_4$.

   Part 1—Standard Stock Solutions: Prepare stock solutions using the table below. Store in brown glass bottles or centrifuge tubes in the dark for up to 3 days.

Concentration	Volume of $KMnO_4$ stock solution	Volume of deionized water
0.005 M	0.250 mL	9.750 mL
0.01 M	0.50 mL	9.50 mL
0.015 M	0.750 mL	9.250 mL
0.02 M	1.00 mL	9.00 mL

Part 2—Dilution Step: To prepare working standards, mix 0.5 mL of each stock solution with 49.5 mL of deionized water in 50 mL centrifuge tubes. This will yield working standards (0.00005, 0.0001 M, 0.00015, and 0.0002 M $KMnO_4$). Prepare fresh daily.

II. *Sample Preparation*:

Sample preparation consists of two key steps: the sample reaction and the sample dilution. These processes work together to ensure accurate and reliable results, as illustrated in the accompanying example (Fig. 2.1).

Soil and solution standards are prepared similarly to unknown samples. The soil standard, made from a large batch of pulverized and homogenized air-dried soil, serves as a laboratory reference for quality control in POXC analyses across different batches and days.

The solution standard, prepared without soil, also functions as a quality control reference, helping to detect contamination in reagents or lab equipment by oxidizing agents or carbon.

2.4 Sample Reaction

1. Weigh 2.50 grams (± 0.05 g) of air-dried soil into one tube. Prepare a soil standard with 2.50 g of known pulverized soil and set aside the other tubes.
2. Add 18.0 mL of deionized water and 2.0 mL of 0.2 M $KMnO_4$ stock solution to each soil tube.
3. For a solution standard, combine 18.0 mL of deionized water and 2.0 mL of 0.2 M $KMnO_4$ in a tube without soil, processing it like the unknown samples.
4. Place the tubes on the shaker and shake them at 240 oscillations per minute for 2 min.
5. Once the 2 min shaking period is complete, carefully remove the tubes from the shaker. Vigorously swirl or manually shake each tube to dislodge any soil particles that may be adhering to the sides or the cap. To prevent further agitation of the soil as it begins to settle, take off the caps at this stage. Place the tubes in a dark, undisturbed area, and allow the soil to settle for approximately 10 min.

2.5 Sample Dilution

1. While the soil samples settle at room temperature, carefully measure and add 49.5 mL of deionized water into each of the second set of centrifuge tubes, ensuring that the water is free from impurities.
2. After allowing the samples to settle for 10 min, swiftly and gently transfer 0.5 mL of the clear supernatant being cautious to avoid any visible particulates or sediments into the tube containing the 49.5 mL of deionized water. It is important to work quickly during this step, as the permanganate solution will continue to react with the soil components.
3. Once the supernatant is added, securely cap each tube and gently invert them several times to ensure thorough mixing of the solutions. These final mixed solutions will maintain their stability for up to 24 h, provided they are stored in a dark environment to prevent any light-sensitive reactions.

2.6 Reading Samples on Spectrophotometer

This method works well on both single cuvette machines and 96-well plate spectrophotometers.

1. Clear polystyrene flat-bottom cell culture plates are sufficient; expensive UV-transparent plates are unnecessary.
2. Replicate all standards and deionized water blanks, running each three or more times for reliable results.
3. Record the absorbance of standards at 550 nm using the spectrophotometer.
4. Subtract the average absorbance of the deionized water blanks from all values. The standard curve's intercept should be close to zero (Flowchart 2.1).

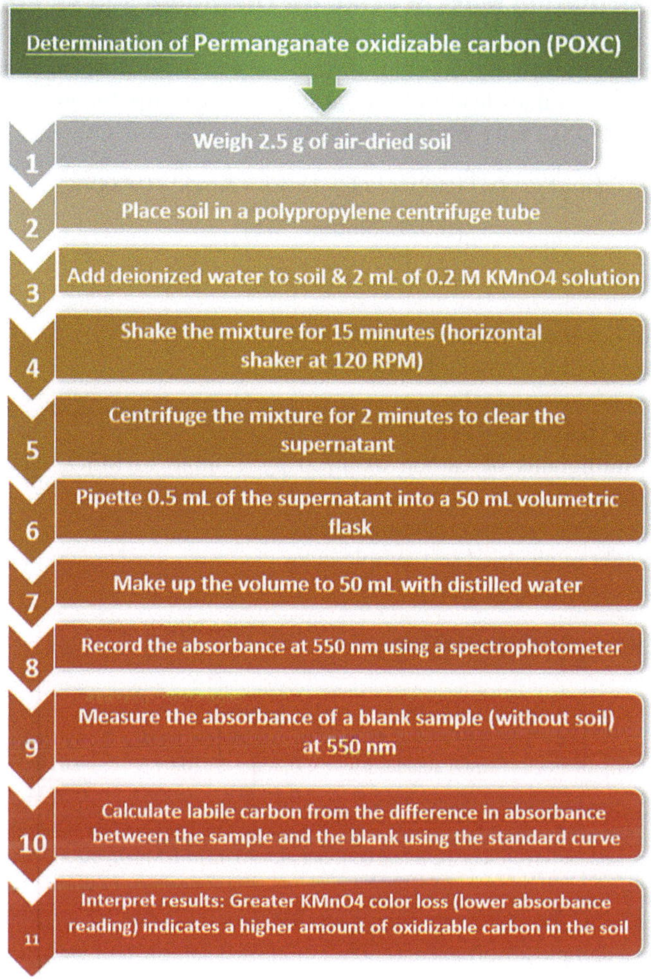

Flowchart 2.1 Determination of permanganate oxidizable carbon (POXC)

2.7 Calculating Mass of POXC for Unknown Soil Samples

1. The amount of carbon oxidized is directly related to the permanganate reduced; higher POXC values lead to lower absorbance. Use the following equation to determine POXC, after Weil et al. (2003):

$$\text{POXC}\left(\text{mgkg}^{-1}\right)=\left[0.02\text{mol}/\text{L}-\left(a+b\times\text{Abs}\right)\right]\times\left(9000\text{mgC}/\text{mol}\right)\times\left(0.02\text{Lsolution}/\text{Wt}\right)$$

where:

- 0.02 mol/L = initial solution concentration
- a = intercept of the standard curve
- b = slope of the standard curve
- Abs = absorbance of unknown
- 9000 = milligrams of carbon oxidized by 1 mole of MnO_4 changing from Mn^{7+} to Mn^{2+}
- 0.02 L = volume of stock solution reacted
- Wt = weight of air-dried soil sample in kg

2.8 Example Calculation

Construct standard curve with the following values:

Y-axis (Molarity of stock $KMnO_4$ standards)	0.005	0.01	0.015	0.02
X-axis (Abs values from spectrophotometer)	0.1000	0.1984	0.3034	0.3966

Note: The standard curve should use the molarity of the stock standards, and not the working standards, since the stock standards represent the actual concentration (0.02 M $KMnO_4$) used to react with the soil

This produces the regression line: $y = 0.0502x - 0.00004$; $R2 = 0.999$
Unknown sample absorbance: 0.3087; unknown sample soil weight: 2.48 grams

$$\text{POXC}\left(\text{mgkg}^{-1}\text{soil}\right)=\left[0.02\text{M}-(-0.00004+\left(0.0502\times0.3087\right)\right]\times\left(9000\text{mgC}/\text{mol}\right)\times\left(0.02\text{Lsolution}/0.00248\text{kg}\right)=329.75\text{mgPOXCkg}^{-1}\text{soil}$$

2.9 Clean-Up and Disposal

Leave the capped centrifuge tubes on the benchtop for a week to allow the permanganate to react completely with the soil, losing its purple color. Afterward, the liquid can be disposed of down the sink, while the soil tubes can be discarded or cleaned for reuse.

Suggested Reading

Weil RR, Islam KR, Stine MA, Gruver JB, Samson-Liebig SE (2003) Estimating active carbon for soil quality assessment: a simplified method for laboratory and field use. Am J Altern Agric 18:3–17

Water-Soluble Carbon in Soils (WSC) 3

Abstract

Water-soluble carbon (WSC), the most mobile and bioavailable fraction of soil organic carbon, is extracted using deionized water, and its concentration is determined through oxidation with potassium dichromate in the presence of sulfuric acid, followed by titration with ferrous ammonium sulfate. Measuring WSC provides insights into the immediate carbon supply for microbial processes and its implications for soil fertility and sustainability.

Keywords

Water-soluble carbon (WSC) · Soil organic carbon · Bioavailable carbon · Potassium dichromate oxidation · Microbial carbon pool · Soil quality indicator · Carbon cycling

3.1 Introduction

The water-soluble carbon (WSC) in soils is organic carbon's most mobile and bioavailable fraction. It consists of organic compounds easily washed away and accessible to soil microorganisms. WSC is crucial for soil nutrient cycling, microbial activity, and overall health. It is sensitive to changes in land use, management practices, and environmental conditions, making it a valuable indicator for assessing soil quality and ecosystem functioning. Measuring WSC provides insights into the immediate carbon supply available for microbial processes and its potential impact on soil fertility and sustainability.

S. Kale et al., *Soil Carbon Fractionation*, Sustainability Sciences in Asia and Africa,
https://doi.org/10.1007/978-981-95-4068-6_3

3.2 Principle

The soil's water-soluble amount of organic carbon is extracted using deionized water. Carbon dissolved in the extract is measured against the amount of potassium dichromate consumed as it reacts with the carbon in the presence of sulfuric acid. Any excess potassium dichromate not reduced in that reaction is then measured by titration with standard ferrous ammonium sulfate (FAS) in the presence of a diphenylamine indicator, as described in the process for determining quickly oxidizable organic carbon.

3.3 Reagents/Materials Required

- 0.1 N $K_2Cr_2O_7$: Dissolve 4.904 g potassium dichromate in 1 L of distilled water.
- Conc. H_2SO_4 and Orthophosphoric acid.
- Ferroin indicator: Dissolve 0.695 g of $FeSO_4 \cdot 7\ H_2O$ and 1.485 g of orthophenanthroline monohydrate in 100 mL distilled water.
- Prepare 0.01 N Ferrous ammonium sulfate solution by dissolving 3.92 g of FAS in a volumetric flask with distilled water. Add 2 mL of sulfuric acid and then make the volume up to 1 L with distilled water.
- Centrifuge tube.

3.4 Procedure

1. Weigh 10 g of air-dried soil and place it into a centrifuge tube. Then, add 20 mL of distilled water. Shake the tube on a horizontal shaker for an hour, followed by centrifugation at 6000 RPM for 15 min until the liquid above the sediment is clear. If necessary, filter the clear liquid (supernatant) and pipette a 10 mL sample into a 250 mL conical flask.
2. In the conical flask, add 2 mL of 0.1 N $K_2Cr_2O_7$, 10 mL of concentrated sulfuric acid, and 5 mL of orthophosphoric acid. Heat the conical flask in a water bath at 100 °C for half an hour.
3. Cool the sample and titrate with freshly prepared 0.01 N FAS using 2–3 drops of ferroin indicator.
4. Also, run a blank without soil extract (Flowchart 3.1).

Flowchart 3.1 Determination of water-soluble carbon

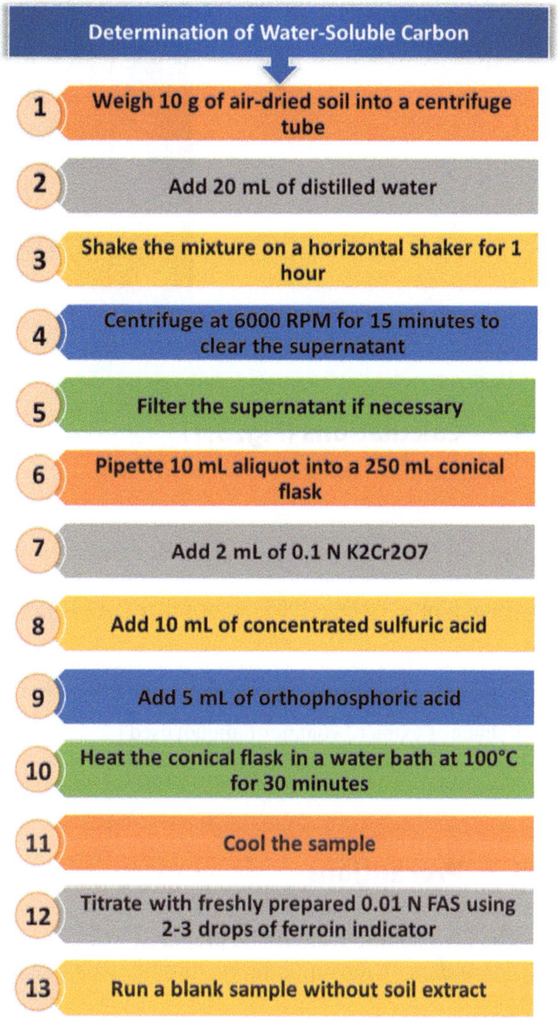

Fig. 3.1 Analysis of water-soluble carbon

3.5 Calculations (Fig. 3.1)

$$\text{Water soluble carbon}\left(\text{mg kg}^{-1}\right) = \left(\text{Blank} - \text{titrated}\right) \times \left(\begin{array}{l}\text{volume of } K_2Cr_2O_7 \text{ taken} \times N \text{ of} \\ K_2Cr_2O_7 \text{ / blank reading of FAS}\end{array}\right)$$

$$\times \text{Normality of } K_2Cr_2O_7 \times 3000 / \text{wt of soil}\left(\text{g}\right) \times \left(\text{Volume of extract / volume of aliquot used}\right)$$

Or

$$\text{WSC}\left(\text{mg kg}^{-1}\right) = \left(\text{Blank} - \text{titrated}\right) \times N \text{ of FAS} \times N\ K_2Cr_2O_7 \times 3000 / \text{wt of soil}\left(\text{g}\right)$$

$$\times \left(\text{Volume of extract / volume of aliquot used}\right)$$

3.6 Precautions

1. The FAS solution should be prepared fresh.
2. The indicator ferroin is more suitable when dealing with soil with low organic carbon content. In this case, the solution's color will change from green to bluish-green and finally to greyish-red.

Suggested Reading

Tao S, Lin B (2000) Water soluble organic carbon and its measurement in soil and sediments. Water Res 34(5):1751–1755

Soil Microbial Biomass Carbon (SMBC) 4

Abstract

Soil microbial biomass carbon (SMBC), a key determinant of nutrient cycling and carbon turnover, is estimated using the fumigation–extraction method, where the difference in organic carbon between fumigated and non-fumigated soil reflects the living microbial biomass. This measurement provides valuable insights into soil biological activity and its role in sustaining ecosystem productivity.

Keywords

Soil microbial biomass carbon (SMBC) · Fumigation-extraction method · Active microbial carbon · Chloroform fumigation · Soil microbial activity · Soil carbon turnover · Nutrient cycling

4.1 Introduction

Soil contains living microbes essential for nutrient transformation, biomass decomposition, and nutrient cycling. The soil microbial biomass carbon, along with climate, controls carbon turnover and nutrient cycling. Living microbial biomass is crucial for plant residue decomposition and the production of plant nutrients.

4.2 Principle

SMBC is determined by the fumigation-extraction method in an incubated soil sample at 27 °C for 5 days at field capacity. Duplicate fresh soil samples are subjected to carbon estimation with and without fumigation with chloroform. The difference in soil organic carbon with and without fumigation is treated as soil microbial biomass carbon.

S. Kale et al., *Soil Carbon Fractionation*, Sustainability Sciences in Asia and Africa,
https://doi.org/10.1007/978-981-95-4068-6_4

4.3 Reagents/Materials

- 0.2 N $K_2Cr_2O_7$: Dissolve 0.9808 g of $K_2Cr_2O_7$ in 100 mL of distilled water.
- Ferroin indicator: Dissolve 0.695 g of $FeSO_4{\cdot}7\ H_2O$ and 1.485 g of orthophenanthroline monohydrate in 100 mL of distilled water.
- 0.5 M K_2SO_4: Dissolve 43.563 g of potassium sulfate in distilled water and dilute to 500 mL.
- M ferrous ammonium sulfate (FAS): Dissolve 3.92 g of FAS in distilled water, add 5–7 drops of concentrated H_2SO_4, and makeup to 1 L.
- For distilled chloroform: Take 20 mL of chloroform for every 10 g of sample, wash it twice with concentrated H_2SO_4, and carefully discard the acidic bottom phase after shaking. To obtain chloroform from ethanol, wash it twice with an equal volume of distilled water and collect the whitest bottom phase or white residue.
- Orthophosphoric acid.
- Conc. $H_2\ SO_4$.
- Desiccator.
- Separating funnel.
- Vacuum pump.

4.4 Equipment/Instruments

1. Rotary shaker
2. Vacuum desiccator and vacuum pump
3. Hot plate

4.5 Procedure

1. Weigh 10 grams of field moist or incubated soil in two 100 mL beakers. If using field moist samples, record the amount of water in the soil for accurate calculations.
2. Put about 40 mL of ethanol-free chloroform and some glass beads in a 100-mL beaker and place it in a vacuum desiccator.
3. Line the inside of the desiccator with filter paper.
4. Keep one set of samples for fumigation in the vacuum desiccator and make sure the desiccator is properly sealed by putting petroleum jelly on the lid.
5. Turn on the vacuum pump and let the chloroform boil for about 5 min.
6. Close the desiccator outlet and store it in a dark place for 24 h.
7. After 24 h, release the vacuum, remove the beaker with chloroform, and remove the filter paper.
8. Suck out any extra chloroform vapors inside the desiccator to make sure they are removed.
9. Let the vacuum out slowly.

10. Take the fumigated samples from the desiccator and the unfumigated samples from the refrigerator.
11. Put these samples in a 250 mL flask and add 25 mL of 0.5 M potassium sulfate (K_2SO_4).
12. Shake the flask for 30 min and then filter the contents.
13. Put 10 mL of the liquid in a 250 mL flask. Add 2 mL of potassium dichromate (0.2 N), 10 mL of concentrated sulfuric acid, and 5 mL of orthophosphoric acid to each flask.
14. Do at least two blank tests with 10 mL of distilled water simultaneously.
15. Put the flasks on a water bath or hot plate at 100 °C for 30 min.
16. Then, add 25 mL of distilled water to each flask and let them cool down to room temperature.
17. Add 2–3 drops of ferroin indicator and test against standard 0.01 N ferrous ammonium sulfate until it turns brick red (Flowchart 4.1).

4.6 Calculations

1. Percent moisture in sample = (wt. of fresh soil – wt of dry soil) × 100/wt of dry soil (g)
2. Net soil weight taken (g) = Wt. of wet soil × 100/(100 + per cent moisture)
3. Volume of extract = Wt. of moist soil – wt. dry soil + vol. of extractant added
4. Normality of FAS = vol. of pot. Dichromate × strength of potassium dichromate/vol. of FAS consumed for blank
5. Volume of potassium dichromate consumed by FAS in sample = Normality of FAS × Vol. of FAS/normality of potassium dichromate
6. Volume of potassium dichromate consumed = 2 – Vol. of potassium dichromate consumed by FAS in sample
7. Extractable carbon (μg/mL) = 0.003 × 0.2 × 10^6 [2 – Vol. of potassium dichromate consumed by FAS in sample]/Vol. of aliquot taken
8. Total carbon in fumigated and unfumigated samples = Extractable carbon (μg/mL) × Volume of extract/Net soil weight taken (g)
9. MBC (μg/g) = (Total carbon in the fumigated sample – total carbon in unfumigated sample)/efficiency factor, i.e., (0.5)

4.7 Precautions

1. Do not air-dry soils before measuring biomass.
2. Pre-incubate soil before measuring biomass for better results.
3. Ensure CHC13 is alcohol-free.
4. This method is unsuitable for soils with recent additions to organic matter.
5. Change the vacuum pump oil more frequently than usual because the oil traps chloroform (Fig. 4.1).

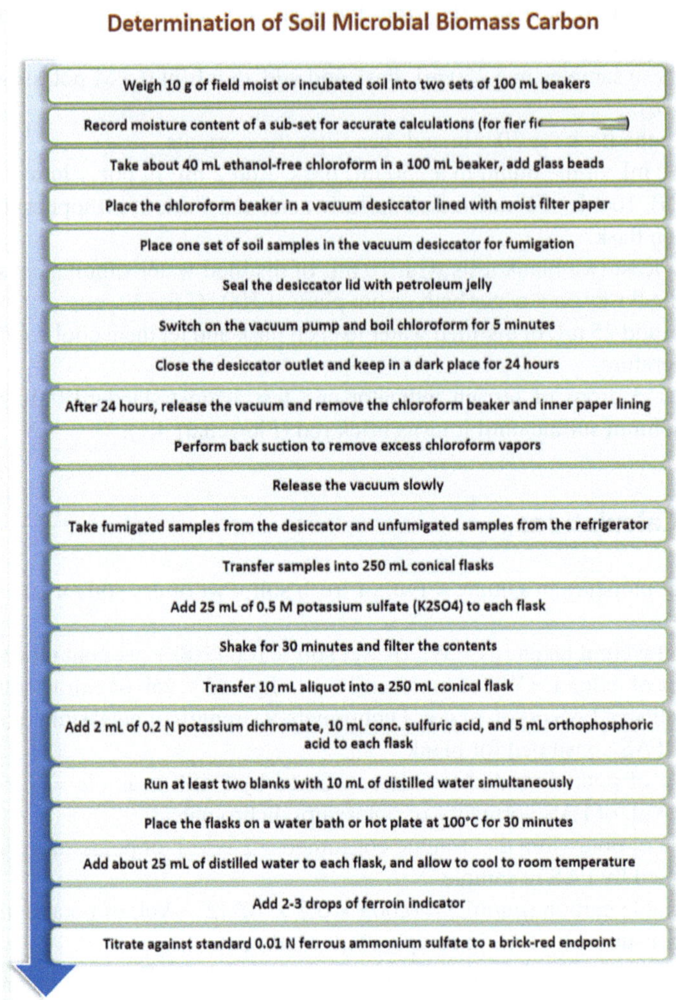

Flowchart 4.1 Determination of soil microbial biomass carbon

Fig. 4.1 Setup for measuring SMBC using chloroform fumigation extraction technique (CRAL laboratory, ICRISAT)

Suggested Reading

Jenkinson DS, Powlson DS (1976) The effects of biological treatments on metabolism in soil. V. A method for measuring soil biomass. Soil Biol Biochem 8:209–213

Vance ED, Brookes PC, Jenkinson DS (1987) An extraction method for measuring soil microbial biomass. Soil Biol Biochem 19:703–707

Light Fraction (LF) Carbon

5

Abstract

Light fraction carbon (LFC), a physically distinct and relatively unstable component of soil organic matter, consists of partially decomposed residues and serves as a sensitive indicator of management impacts on soil health, isolated through density fractionation techniques. This approach provides valuable insights into the labile carbon pool and its role in nutrient cycling and soil sustainability.

Keywords

Light fraction carbon · Soil organic matter fractionation · Density separation · Sodium iodide method · Labile carbon pool · Soil quality indicator · Soil carbon dynamics

5.1 Introduction

The light fraction material mainly consists of organic residues that have not completely decomposed. It is a poorly decomposed part of SOM that is relatively unstable and contains easily decomposable components like decomposed carbohydrates. It is mainly made up of free organic matter, which is either partially decomposed or not completely decomposed. The light fraction carbon represents the physically based SOM pool and can be a sensitive indicator of how different farming practices affect soil organic matter.

S. Kale et al., *Soil Carbon Fractionation*, Sustainability Sciences in Asia and Africa,
https://doi.org/10.1007/978-981-95-4068-6_5

5.2 Principle

Isolation of LF from soils involves densitometric techniques. To separate the light fraction carbon, the soil is suspended in a dense liquid, and the material that floats is considered the light fraction carbon. Various methods can be used to isolate the light fraction carbon from the entire soil or aggregates of various sizes. It includes stirring or centrifugation of soil, suction of light fraction from the soil after equilibrium is achieved, soil dispersion by low or high sonication or using sodium hexametaphosphate, equilibrium time, etc. Density fractionation may be accomplished with cheaper NaI salts, which may be used when many samples need to be analyzed. However, other salts, such as sodium poly tungstate, molybdenum poly tungstate, etc., may also be used.

5.3 Reagents/Materials

- *Sodium Polytungstate Solution*: Dissolve 185 g of sodium polytungstate in 100 mL of distilled water.
- *Sodium Iodide Solution*: Dissolve 170 g of sodium iodide in 100 mL of distilled water.
- *0.2 N Potassium Dichromate Solution*: 4.904 g of potassium dichromate in distilled water and adjust the total volume to 500 mL.
- *M Ferrous Ammonium Sulfate (FAS) Solution*: Dissolve 3.92 g of ferrous ammonium sulfate in distilled water, add 5–10 drops of concentrated sulfuric acid (H_2SO_4), and make up the volume to 1 L.
- *Ferroin Indicator*: Dissolve 0.695 g of ferrous sulfate heptahydrate ($FeSO_4 \cdot 7H_2O$) and 1.485 g of ortho-phenanthroline monohydrate in 100 mL of distilled water.
- *Equipment Required*: Centrifuge tubes, centrifuge, and TOC analyzer.

5.4 Procedure

In a centrifuge tube, suspend 5 g soil in 35 mL of sodium iodide (1.7 g cc^{-1}). Mix the content slowly by giving about ten strokes so that the whole soil is suspended. Keep the centrifuge tube for 20 minutes to remove the entrapped air. Centrifuge the suspension for 60 minutes at 20 °C. Filter or aspirate the floating material on filter paper and rinse thoroughly with distilled water to remove excess salts. Dry the filter paper at 70 °C and collect and weigh the Light Fraction (LF) retained on the filter paper. Sodium polytungstate (SG 1.85 g cc^{-1}) is costlier than NaI; thus, the latter is preferred when many samples are involved. Analyze the total carbon content in the LF collected with the TOC analyzer. Alternatively, follow the procedure described for MBC using potassium dichromate as an oxidizing agent (Flowchart 5.1).

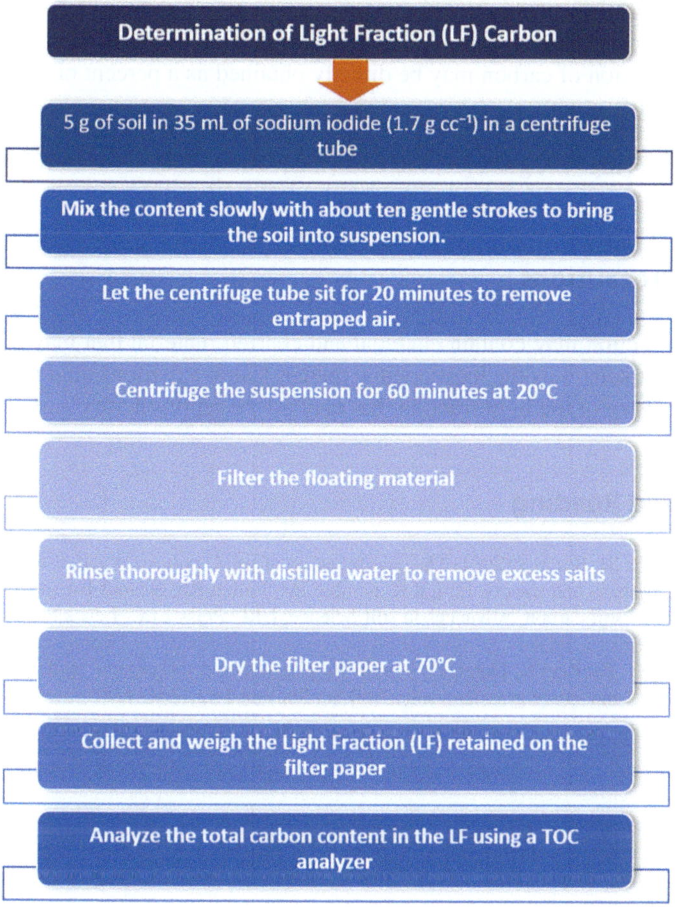

Flowchart 5.1 Determination of light fraction (LF) carbon

5.5 Calculations

The light fraction of carbon may be directly obtained as a percent of the content in the collected fraction.

$$\mathrm{LFC}\left(\mathrm{mg}/100\,\mathrm{g\,soil}\right)=\left[\%\mathrm{C}\times\mathrm{LF}\left(\mathrm{mg}\right)/100\right]\times 100/\,\mathrm{wt\,of\,soil}\left(5\,\mathrm{g}\right)$$

5.6 Precautions

The light fraction may contain charcoal and charred material that may not decompose. In peat soils, it is the inert organic matter. Using a denser liquid may result in higher recovery of the light fraction, but may not be representative.

Suggested Reading

Paul EA, Morris SJ, Bohm S (2001) The determination of soil C pool sizes and turnover rates: biophysical fraction and tracers. In: Lal R, Kimble JM, Follett RF, Stewart BA (eds) Assessment methods for soil carbon, Advances in soil Science. CRC Press LLC, Lewis Publishers, USA, pp 193–206

Six J, Elliott ET, Paustian K, Doran JW (1998) Aggregation and soil organic matter accumulation in cultivated and native grassland soils. Soil Sci Soc Am J 62:1367–1377

Determination of Soil Acid Hydrolysable Carbohydrates

6

Abstract

Soil carbohydrates, constituting 5–25% of soil organic matter, are key indicators of substrate quality and SOM dynamics, typically assessed through acid hydrolysis and chromatography despite extraction challenges due to humic associations. Understanding soil carbohydrate pools provides insights into labile carbon fractions and their contribution to soil health and sustainability.

Keywords

Acid hydrolysable carbohydrates · Soil organic matter · Labile carbon pool · Sulfuric acid hydrolysis · Soil carbon analysis

6.1 Introduction

The composition of soil polysaccharides includes hexoses, methylhexoses, 6-deoxyhexoses, glucose, etc. The pentoses (uronic acids) and sugar alcohols are less sensitive to the anthrone method. In general, the C_{carb}/SOC and C_{mic}/SOC ratios of subtropical soils are lower than those of tropical soils, indicating that the substrate quality of soil organic matter (SOM) is low.

6.2 Principle

The carbohydrates in the soil can make up 5–25% of the soil's organic matter and exist in various forms. Their content and composition are typically determined through acid hydrolysis and chromatography. However, some carbohydrates and proteins may be challenging to extract using acid extraction methods due to their

S. Kale et al., *Soil Carbon Fractionation*, Sustainability Sciences in Asia and Africa,
https://doi.org/10.1007/978-981-95-4068-6_6

association with humic material or inorganic colloids, limiting the assessment of labile soil organic carbon pools.

6.3 Reagents

1. 24 N H_2SO_4
2. 0.2% anthrone
3. Glucose standard
4. 6 N NaOH
5. Colorimeter

6.4 Procedures

1. Weigh out 5 g of oven-dried soil that has been passed through a 150 μm mesh sieve and transfer it into a 150 mL conical flask.
2. Add 10 mL of 24 N sulfuric acid (H_2SO_4) to the soil and let the mixture undergo hydrolysis for 24 hours in a steam bath set at approximately 100 °C.
3. Cover the flask to minimize loss from evaporation during the hydrolysis process.
4. After the hydrolysis is complete, filter the hydrolyzed mixture using a sintered glass funnel (porosity 3). Neutralize the resulting filtrate to a pH of 6.8 by adding a 6 N sodium hydroxide (NaOH) solution. Rinse the remaining residue with 10 mL of hot water for each gram of soil.
5. Let the filtrate cool to room temperature, then dilute it with distilled water in a volumetric flask to a final volume of 50 mL.
6. If a dark-colored precipitate forms, it should be removed by centrifuging the solution at 10,000 rpm for 10 minutes. The clear liquid that remains after centrifugation is the acid hydrolysate.
7. Using a pipette, transfer 5 mL of the hydrolysate into 150 × 25 mm test tubes. Quickly add 10 mL of 0.2% anthrone solution and shake the tube immediately to thoroughly mix the contents. The mixture should turn green.
8. Measure the optical density of the solution at 625 nm using a spectrophotometer.
9. To create a standard curve, prepare glucose solutions with concentrations of 0, 5, 10, 20, 40, and 80 mg L^{-1} in 50 mL volumetric flasks. Take 5 mL of each solution, add 10 mL of anthrone reagent, and measure the absorbance at 625 nm using a spectrophotometer. Plot the percent transmission against the concentration to create the standard curve. Use a blank containing 5 mL of distilled water and 10 mL of anthrone reagent as the reference blank for the acid-anthrone method (Flowchart 6.1).

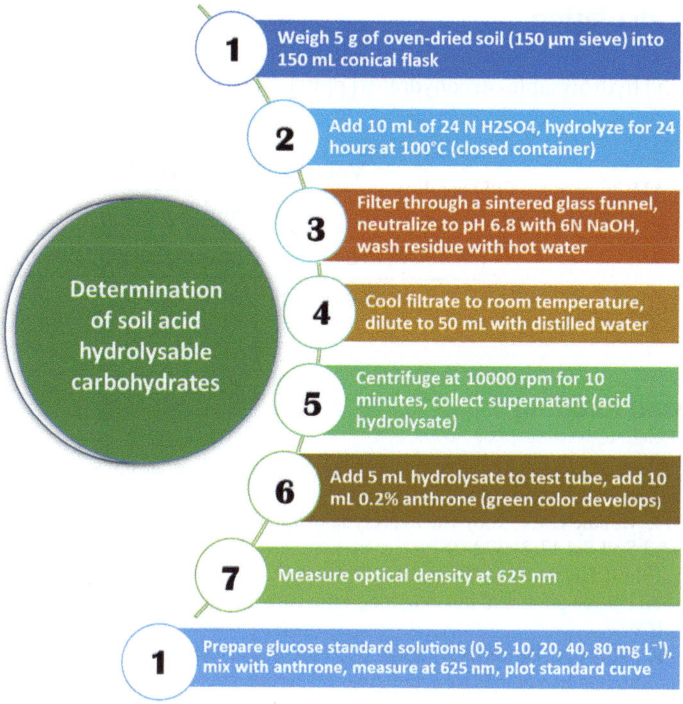

Flowchart 6.1 Determination of soil acid hydrolysable carbohydrates

6.5 Calculation

$$\text{Acid hydrolysable carbohydrates}\left(\text{ppm}\right) = \text{OD}\left(\text{ppm}\right) \times \text{dilution factor}$$

$$\text{AHC}\left(\text{ppm}\right) = \text{O.D.reading} \times \frac{\text{Vol.taken by } 24\,\text{N}\,\text{H}_2\text{SO}_4}{\text{Wt.of soil}\left(\text{g}\right)} \times \frac{\left(\text{Vol.of acid extract} + \text{Anthrone reagent taken}\right)}{\text{Vol.of acid extract taken}}$$

Suggested Reading

Brink RH, Dybar P, Lynch DLN (1960) Measurement of carbohydrates in soil hydrolysates with anthrone. Soil Sci 89:157–166

Cheshire MV, Mundie CM (1966) The hydrolytic extraction of carbohydrates from soil by sulphuric acid. J Soil Sci 17(2):372–381

Collins HP, Elliott ET, Paustian K, Bundy LG, Dick WA, Huggins DR, Smucker AJM, Paul EA (2000) Soil carbon pools and fluxes in long-term corn belt agroecosystems. Soil Biol Biochem 32:157–168

Motavalli PP, Palm CA, Parton WJ, Elliott ET, Frey SD (1994) Comparison of laboratory and modeling simulation methods for estimating soil carbon pools in tropical forest soils. Soil Biol Biochem 26:935–944

Paul EA, Harris D, Collins HP, Schulthess U, Robertson GP (1999) Evolution of CO_2 and soil carbon dynamics in biologically managed, row-crop agroecosystems. Appl Soil Ecol 11:53–65

Labile Pool-I and Labile Pool-II Carbon: Two-Step Acid Hydrolysis

7

Abstract

Labile carbon pools, comprising rapidly and moderately degradable fractions of soil organic carbon, are key indicators of soil quality and carbon dynamics, extracted through sequential acid hydrolysis to assess short- and long-term nutrient cycling. Assessing these pools offers valuable insights into carbon dynamics, soil quality, and resilience in agricultural systems.

Keywords

Labile carbon pool · Acid hydrolysis method · Soil organic carbon fractions · 5 N sulfuric acid extraction · Cellulose decomposition · Short-term nutrient cycling · Soil carbon dynamics · Soil quality indicator

7.1 Introduction

Labile carbon pools-I and II are easily degradable fractions of the soil organic carbon. Labile Pool-I contains rapidly mineralizable compounds, including sugars and proteins, which act as a readily available energy source for the soil microbial population and, therefore, influence the short-term cycling of nutrients (Taylor et al. 2002). Labile Pool-II comprises moderately decomposed material, such as cellulose, which provides sustained release of nutrients and contributes towards long-term soil fertility (Berg and McClaugherty 2008). These pools collectively help

S. Kale et al., *Soil Carbon Fractionation*, Sustainability Sciences in Asia and Africa,
https://doi.org/10.1007/978-981-95-4068-6_7

contribute to those important roles of soil health and productivity in carbon dynamics; as a result, they act as indicators of soil quality and resilience in agricultural systems.

7.2 Principle

The acid hydrolysis method extracts two labile pools: labile pool I (LP-I) using 5 N H_2SO_4 and labile pool II (LP-II) using 26 N H_2SO_4. LP-I contains polysaccharides from plants and microbial sources, while LP-II is primarily composed of cellulose. Recalcitrant C (RP-C) pools are calculated by subtracting the total organic carbon (TOC) from the combined concentration of the labile pools (LP-I and LP-II).

7.3 Procedure

1. The assessment of labile carbon pools was carried out using a two-step acid hydrolysis method following the approach of Rovira and Vallejo (2002), where sulfuric acid (H_2SO_4) was used as the extractant. Initially, 20 mL of 5 N sulfuric acid was mixed with 0.5 g of soil, and the mixture was hydrolyzed in a hot water bath set at 105 °C for 30 minutes using sealed 100 mL centrifuge tubes.
2. After the hydrolysis process, the samples were allowed to cool, then centrifuged at 6000 rpm for 10 minutes. The resulting supernatant was poured off, combined with a 20 mL rinse of deionized water, and filtered using Whatman no. 1 filter paper. The resulting solution was labeled as labile pool I (LP1).
3. The remaining residue underwent a second hydrolysis step for 16 hours at room temperature with continuous shaking, using 2 mL of 26 N sulfuric acid. The following day, the acid was diluted to a concentration of 2 N by adding 24–26 mL of deionized water, and the sample was hydrolyzed again at 105 °C for 3 hours, with intermittent shaking. The recovery procedure for the second hydrolysate, labeled as labile pool II (LP2), was performed similarly to LP1.
4. The carbon content in the hydrolysate was analyzed using the Nelson and Sommers (1982) method. In this method, 4 mL of the hydrolysate was oxidized with 1 mL of 0.066 M potassium dichromate ($K_2Cr_2O_7$) and 5 mL of concentrated sulfuric acid (H_2SO_4) at 150 °C for 30 minutes.
5. Once cooled, the samples were titrated using o-phenanthroline indicator (2–3 drops) against 0.033 M ferrous ammonium sulfate (FAS) until the solution changed color from greenish-violet to brick red.

7.4 Equipment

1. Centrifuge tube
2. Centrifuge
3. Hot water bath
4. Whatman filter paper

7.5 Reagents

1. Conc. H_2SO_4
2. Potassium dichromate ($K_2Cr_2O_7$)
3. O-phenanthrolene
4. Ferrous ammonium sulfate (Flowchart 7.1)

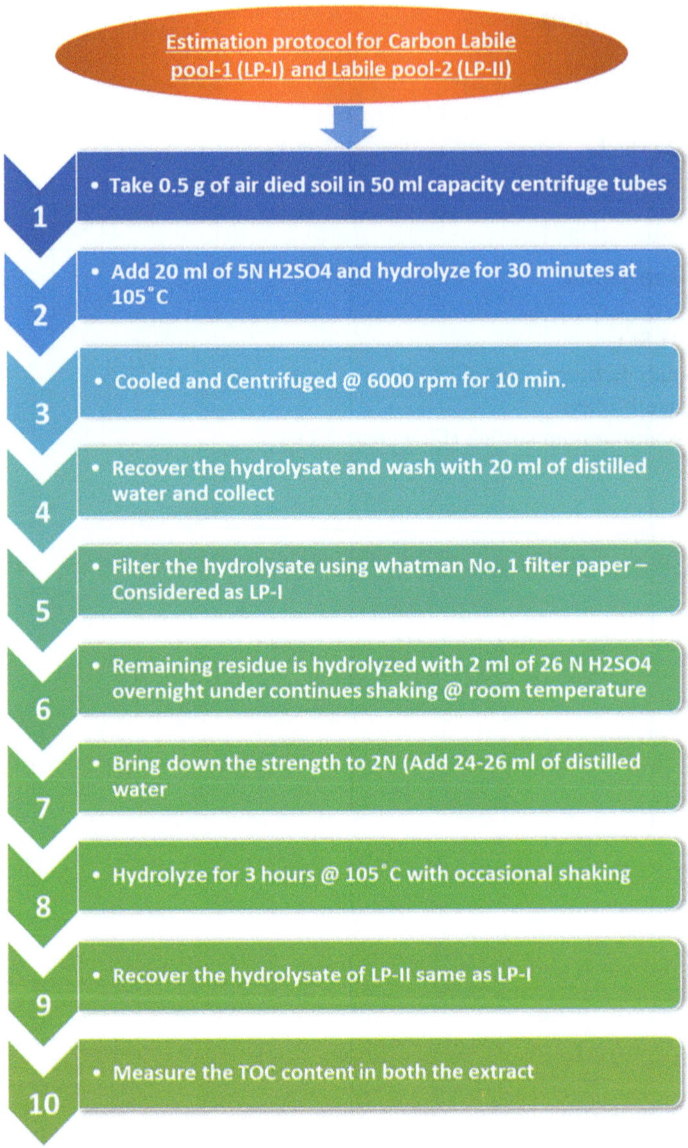

Flowchart 7.1 Estimation protocol for Carbon Labile pool-1 (LP-I) and Labile pool-2 (LP-II)

7.6 Calculation

$$\mathrm{OC}(\%) = \{(A \times M \times 0.003) / g\} \times \{(E / S\} \times 100$$

where

- T = Standardization titer value
- M = Molarity of FAS (approximately 0.033 M)
- A = (mL HB – mL sample) × {(mL UB – mL HB)/mL UB} + (mL HB – mL sample)
- E = Extraction volume (mL)
- g = Weight of dry soil sample (g)
- S = Volume of digest sample (mL)
- M = 0.4/T

Explanation of variables:

- mL HB = Volume of blank titration (mL)
- mL sample = Volume of the sample titration (mL)
- mL UB = Volume of unbuffered blank titration (mL)

This formula calculates the organic carbon (OC) percentage in the soil sample, considering titration volumes, soil mass, extraction, and sample volumes, along with the molarity of ferrous ammonium sulfate (FAS).

Suggested Reading

Berg B, McClaugherty C (2008) Influence of soil and plant community factors. In: Plant litter: decomposition, humus formation, carbon sequestration. Springer Berlin Heidelberg, Berlin, Heidelberg, pp 175–192

Nelson DW, Sommers LB (1982) Total carbon, organic carbon and organic matter. In: Page AL, Miller RH, Keeney DR (eds) Methods of soil analysis. Part 2. American Society of Agronomy, Madison, pp 539–579

Rovira P, Vallejo VR (2002) Labile and recalcitrant pools of carbon and nitrogen in organic matter decomposing at different depths in soil: an acid hydrolysis approach. Geoderma 107:109–141

Taylor JP, Wilson B, Mills MS, Burns RG (2002) Comparison of microbial numbers and enzymatic activities in surface soils and subsoils using various techniques. Soil Biol Biochem 34(3):387–401

Hot Water Extractable Carbon (HWEC) 8

Abstract

The hot water extractable carbon (HWEC) refers to the fraction of soil organic carbon that is soluble in hot water. It serves as an indicator for labile carbon, preserving the easily decomposable portion of organic matter. This labile carbon can be readily decomposed by soil microorganisms and plays a crucial role in nutrient cycling and microbial activity in the soil (Strosser 2010). HWEC is primarily used as a key indicator of soil health and quality, and it also reflects the dynamic changes in the organic carbon content of the soil in response to management practices.

Keywords

Hot water extractable carbon (HWEC) · Labile soil organic carbon · Soil health indicator · Potassium dichromate oxidation · Nutrient cycling · Microbial carbon availability · Organic matter decomposition

8.1 Introduction

The hot water extractable carbon (HWEC) refers to the fraction of soil organic carbon that is soluble in hot water. It serves as an indicator for labile carbon, preserving the easily decomposable portion of organic matter. This labile carbon can be readily decomposed by soil microorganisms and plays a crucial role in nutrient cycling and microbial activity in the soil. HWEC is primarily used as a key indicator of soil health and quality, and it also reflects the dynamic changes in the organic carbon content of the soil in response to management practices.

S. Kale et al., *Soil Carbon Fractionation*, Sustainability Sciences in Asia and Africa,
https://doi.org/10.1007/978-981-95-4068-6_8

8.2 Principle

The hot water extractable carbon (HWEC) method quantifies the labile fraction of soil organic carbon soluble in hot water, reflecting the readily decomposable portion of organic matter. An air-dried soil sample is extracted with water and then heated at 80 °C for 16 hours. After filtration, the carbon content in the extract is determined by oxidation with $K_2Cr_2O_7$ and H_2SO_4, followed by titration with ferrous ammonium sulfate, using an o-phenanthroline indicator to detect the endpoint (Ghani et al. 2003).

8.3 Procedure

1. Place *3 g of air-dried soil* into a 50 mL centrifuge tube. Add *30 mL of deionized water* and shake the mixture at *30 rpm for 30 minutes* at room temperature. Centrifuge the suspension at *3000 rpm for 20 minutes* and discard the supernatant to remove cold-water soluble carbon.
2. Add *30 mL of deionized water* to the remaining residue and place the tube in a hot water bath at *80 °C for 16 hours*. After cooling, shake and centrifuge the tubes again at *3000 rpm for 20 minutes*.
3. Filter the supernatant through a *0.45 μm cellulose nitrate membrane*. Measure the carbon concentration in the filtrate using the Nelson and Sommers (1982) *method*.
4. Oxidize a *4 mL aliquot* of the sample with *1 mL of 0.066 M potassium dichromate* ($K_2Cr_2O_7$) and *5 mL of concentrated sulfuric acid* (H_2SO_4) at *150 °C for 30 minutes*. After cooling, titrate the sample with *0.033 M ferrous ammonium sulfate* using *2–3 drops of o-phenanthroline indicator* until the solution turns from greenish violet to brick red (Flowchart 8.1).

8.4 Calculation

$$\text{OC}(\%) = \{(A \times M \times 0.003)/g\} \times \{(E/S\} \times 100$$

where

- T = Standardization titer value
- M = Molarity of FAS (approximately 0.033 M)
- A = (mL HB − mL sample) × {(mL UB − mL HB)/mL UB} + (mL HB − mL sample)
- E = Extraction volume (mL)
- g = Weight of dry soil sample (g)
- S = Volume of digest sample (mL)
- M = 0.4/T

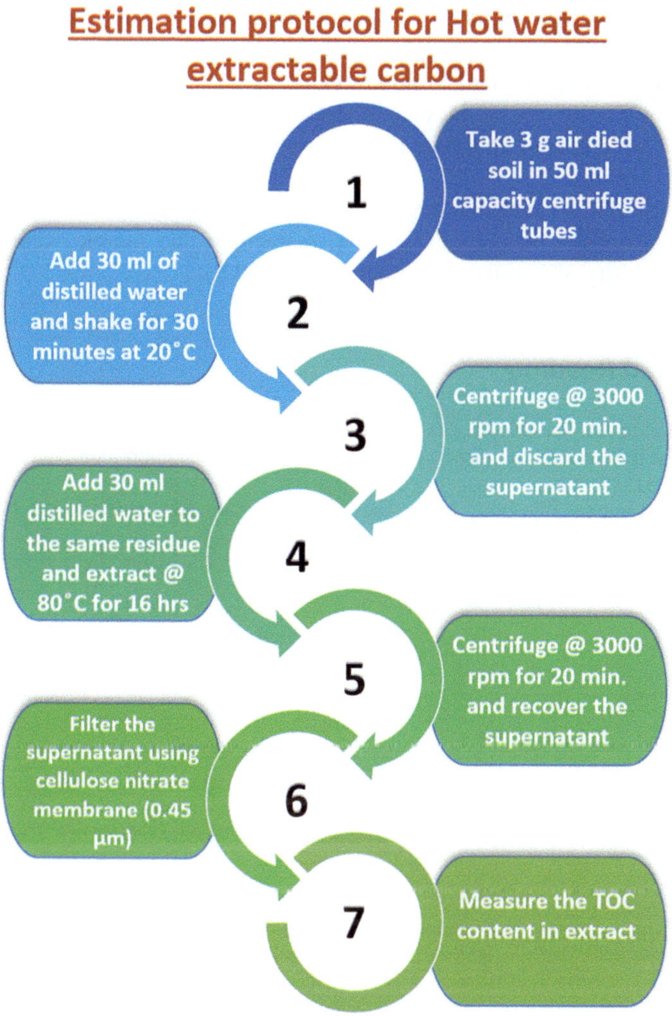

Flowchart 8.1 Estimation protocol for hot water extractable carbon

Explanation of variables:

- mL HB = Volume of blank titration (mL)
- mL sample = Volume of the sample titration (mL)
- mL UB = Volume of unbuffered blank titration (mL)

This formula calculates the organic carbon (OC) percentage in the soil sample, considering titration volumes, soil mass, extraction, and sample volumes, along with the molarity of ferrous ammonium sulfate (FAS).

Suggested Reading

Ghani A, Dexter M, Perrott KW (2003) Hot-water extractable carbon in soils: a sensitive measurement for determining impacts of fertilisation, grazing and cultivation. Soil Biol Biochem 35(9):1231–1243

Nelson DW, Sommers LE (1982) Total carbon, organic carbon, and organic matter. Methods Soil Anal Part 2 9:539–579

Strosser E (2010) Methods for determination of labile soil organic matter: an overview. J Agrobiol 27(2):49

9 Particulate Organic Matter-Carbon and Mineral Associated Organic Carbon

Abstract

Particulate organic matter carbon (POM-C) and mineral-associated organic carbon (MAOC) represent contrasting soil carbon fractions POM-C as a labile, short-term nutrient source and MAOC as a stable, long-term carbon reservoir critical for evaluating soil health and management impacts. Understanding the balance between these fractions provides critical insights into soil carbon storage, stability, and response to management practices.

Keywords

Particulate organic matter carbon (POM-C) · Mineral-associated organic carbon (MAOC) · Soil carbon fractionation · Labile carbon pool · Stable carbon pool · Sodium hexametaphosphate dispersion · Soil health assessment

9.1 Introduction

Particulate Organic Matter-Carbon (POM-C) and Mineral associated Organic C (MAOC) are two very different fractions of soil organic carbon, with implications for assessing soil health and nutrient dynamics. Particulate organic matter carbon consists of relatively large, labile organic particles loosely bound within the soil matrix and readily available for microbial decomposition. It is an indicator of carbon turnover and nutrient availability on a short-term basis. On the other hand, MAOC represents the fraction of organic carbon tightly bound into the soil minerals, thus furnishing a more stable and longer-term carbon reservoir. Understanding this balance between the two fractions shall help evaluate soil carbon storage and its response to management practices.

S. Kale et al., *Soil Carbon Fractionation*, Sustainability Sciences in Asia and Africa,
https://doi.org/10.1007/978-981-95-4068-6_9

9.2 Principle

Particulate organic matter (POM-C) is isolated by combining a specific amount of soil with a dispersing agent, followed by shaking. The mixture is then passed through a *53 µm sieve*, and water is used to wash away silt and clay particles. The material remaining on the sieve, which includes POM and sand, is subsequently analyzed for carbon content.

9.3 Reagents

- Sodium hexametaphosphate

9.4 Equipment

- Reciprocal shaker
- TOC analyzer

9.5 Procedure

1. Weigh 10 g of soil in a conical flask and add 30 mL of sodium hexametaphosphate.
2. Shake the mixture for 15 hours on a reciprocal shaker.
3. Pass the contents through a 53 µm sieve and rinse thoroughly with water to remove silt and clay fractions.
4. Collect the POM plus sand retained on the sieve in a petri dish or beaker and oven-dry at 50 °C.
5. Grind the contents and pass through a 180 µm screen.
6. Analyze the POM plus sand fraction for total carbon using a TOC analyzer.
7. Alternatively, you can determine the carbon content in the POM plus sand fraction as outlined under light fraction (LF) carbon estimation. Make sure to remove all identifiable plant residues before estimation (Flowchart 9.1 and Fig. 9.1).

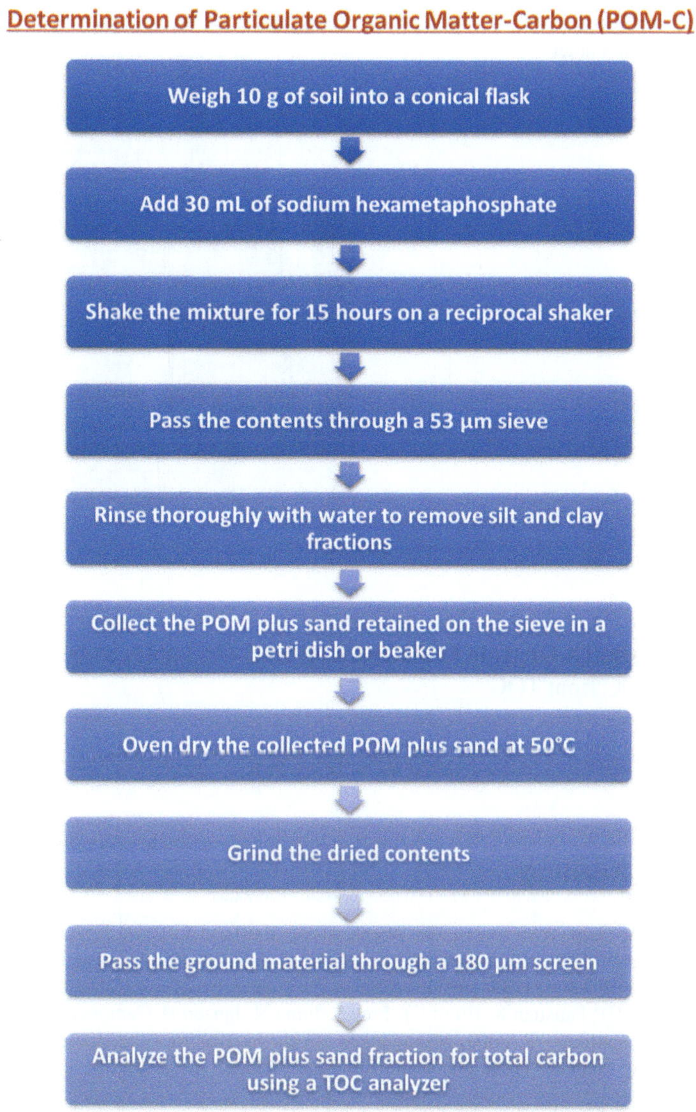

Flowchart 9.1 Determination of particulate organic matter-carbon (POM-C)

Fig. 9.1 Estimation of particulate organic matter carbon

Mineral-associated Organic C (MOAC) MOAC (<53 μm) can be calculated by subtracting POC from TOC.

$$\text{MAOC} = \text{TOC-POC}$$

Suggested Reading

Camberdella CA, Gajda AM, Doran JW, Wienhold BJ, Kettler TA (2001) Estimating particulate and total organic matter by weight loss-on–ignition. In: Advances in soil science. CRC Press LLC, Lewis Publishers, USA, pp 349–359

Paul EA, Collins HP, Paustian K, Elliot ET, Frey S, Juma N, Janzen H, Campbell CA, Zentner RP, Lafond GP, Moulinm AP (2003) Management effects on dynamics and storae rates of organic matter in long–term crop rotations. Can J Soil Sci 83:49–61

Determination of Water-Stable Aggregates Carbon

10

Abstract

The determination of carbon associated with water-stable aggregates is a critical indicator of soil structural stability and long-term carbon sequestration in agricultural and natural ecosystems. This method quantifies organic carbon retained within water-stable aggregates, reflecting the physical protection of carbon against rapid decomposition and erosion.

Keywords

Water-stable aggregates · Aggregate-associated carbon · Soil organic carbon pools · Sand-free carbon · Wet sieving method · Soil aggregate fractionation · Soil structural stability

10.1 Introduction

The organic carbon found within soil aggregates is considered a slow pool of organic carbon. This carbon pool has a mean residence time of 20–250 years and is relatively resistant to microbial decomposition compared to active carbon pools, but much more susceptible than passive carbon pools. The size of the carbon fraction ranges from 2000 to 53 μm. In aggregates, the carbon located between two aggregates is referred to as inter-particulate organic carbon (the light fraction of organic matter carbon), while the carbon within two aggregates is known as intra-particulate organic carbon (the heavy fraction of organic matter carbon). The light fraction of organic carbon can be easily estimated using density fractionation with sodium polytungstate at a density of 1.85/2.0 g cm^{-3}. The heavy fraction is estimated after dispersion and filtration with a 53 μm sieve.

S. Kale et al., *Soil Carbon Fractionation*, Sustainability Sciences in Asia and Africa,
https://doi.org/10.1007/978-981-95-4068-6_10

10.2 Principle

To investigate the slow pools of carbon and nitrogen concentration and mineralization in macro- and micro-aggregates, surface sub-samples (>2 mm) are utilized for aggregate separation through wet sieving (Camberdella and Elliott 1992).

10.3 Instruments

1. Hydrometer
2. Sieves (2000, 250, and 53 μm size classes and pan)
3. Yoder's apparatus
4. Sodium iodide NaI
5. Vacuum drier
6. Ultrasonicator
7. Sodium hexametaphosphate

10.4 Procedure

1. Take a 100 g air-dried soil sample with aggregates larger than 2000 μm on a 2000 μm sieve.
2. Pre-wet the air-dried samples using Yoder's apparatus for 5 minutes at room temperature with deionized water by capillary action.
3. Keep the pan below the 2000 μm sieve to collect the soil.
4. Sieve the soils for 5 minutes inside water by gently moving the sieve vertically (3 cm to 50 times) through a shallow pan.
5. After the 5-minute cycle, gently backwash the stable aggregates (>2000 μm) into an aluminum pan. Discard any floating organic materials (>2 μm) and collect the remaining aggregate size class in the pan (<2000 μm).
6. Transfer the soil passing through a particular sieve and remaining in the shallow pan to the next smaller-size sieve and repeat the process.
7. Pour the water plus soil that goes through the sieve to the pan (<2000 μm) onto the next sieve (250 μm). Keep the pan below this sieve, and repeat the sieving process.
8. After sieving, gently backwash off and collect the soil content in the >250 μm aggregate into an aluminum pan. This aggregate size class is 2000–250 μm.
9. Pour the water plus soil that goes through the sieve to the pan (<250 μm) onto the next sieve (53 μm). Keep the pan below this sieve, and repeat the sieving process.
10. Gently backwash off and collect the soil content in the >53 μm aggregate into an aluminum pan. This aggregate size class is 250–53 μm.

11. Collect the content of soil and water that goes into the pan onto an aluminum pan. Allow it to settle for 24 hours, then decant the water content and allow it to dry at 65 °C in a vacuum drier.
12. Keep the organic carbon estimation from each size class for drying at 65 °C.
13. Quantify the amount of soil present in each aggregate size class, such as 2000, 2000–250, 250–53, and <53 μm.
14. Allow soil aggregates smaller than the 0.53 μm (silt + clay) particle size fraction to settle overnight, then centrifuge at 10000 rpm for 15 minutes, and oven-dry at 65 °C.
15. Grind the dried soil to pass through a 150 μm sieve before chemical analysis.

10.5 Analysis of Sand-Free Carbon from Each Size Class

To determine the percentage of free in the various size classes, the soil sample is shaken overnight with a 1% (W/V) solution of sodium hexametaphosphate using the method outlined by Elliott (1986). The sand fraction can be easily separated from the different size classes by using a combination of 5 g of soil and 15 mL of a 5 g L^{-1} solution of sodium hexametaphosphate, shaking it with a reciprocal shaker overnight, and then sieving it through a 2000 μm sieve. An ultrasonicator can be used to disperse sand from soil aggregates, making the process more efficient and saving chemicals. After the dispersion of soil aggregates, collect the sand in a 53 μm sieve with repeated washing by tap water, quantify the amount of sand and soil, and express on a dry weight basis at 105 °C temperature. Determine the carbon using the wet digestion method and the Kjeldahl distillation method for total nitrogen. These C and N are called aggregate's sand-free carbon and nitrogen (Flowchart 10.1).

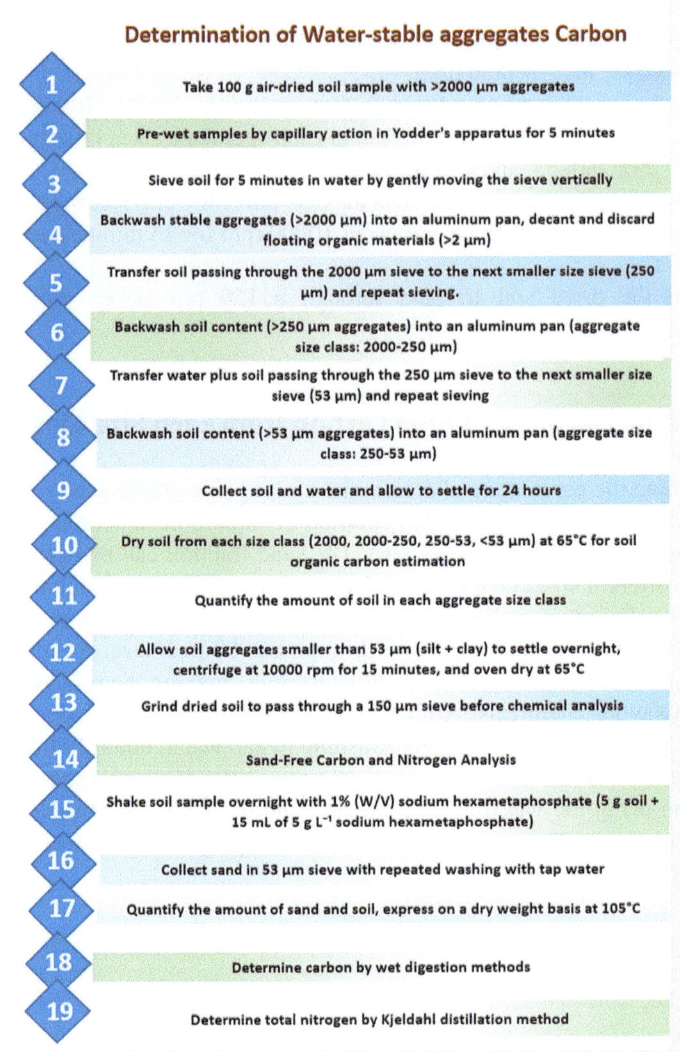

Flowchart 10.1 Determination of water-stable aggregates carbon

10.6 Calculation

The formula for calculating sand-free total C or N concentration is as follows (Six et al. 2000).

$$\text{Sand}-\text{free}\left(\text{C or N}\right)_{\text{fraction}}\ \frac{\left(\text{C or N}\right)_{\text{fraction}}}{1-\left(\text{Sand portion}\right)_{\text{fraction}}}$$

Suggested Reading

Camberdella CA, Elliott ET (1992) Particulate soil organic matter changes across a grassland cultivation sequence. Soil Sci Soc Am J 56:777–783

Camberdella CA, Elliott ET (1993) Carbon and nitrogen distribution in aggregates from cultivated and native grassland soils. Soil Sci Soc Am J 57:1071–1076

Elliott E (1986) Aggregate structure and carbon, nitrogen, and phosphorus in native and cultivated soils. Soil Sci Soc Am J 50:627–633

Six J, Merckx R, Kimpe K, Paustian K, Elliott ET (2000) A re-evaluation of the enriched labile soil organic matter fraction. Eur J Soil Sci 51:283–293

Humic Acid (HA) and Fulvic Acid (FA) Carbon

11

Abstract

Humic acid (HA) and fulvic acid (FA), key passive fractions of soil organic matter, differ in molecular size and solubility, contributing to long-term carbon sequestration and nutrient exchange, and are extracted using base–acid fractionation. Their characterization provides insights into soil quality, carbon stability, and nutrient availability.

Keywords

Humic acid carbon (HA-C) · Fulvic acid carbon (FA-C) · Passive carbon pool · Soil organic matter · Base-acid extraction method · Walkley and Black method · Soil carbon sequestration · Soil health indicator

11.1 Introduction

Humic acid (HA) and fulvic acid (FA) are two of the key components of soil organic matter, which makes an essential contribution to soil fertility and carbon storage. HA is one of the larger and more complex molecules, which are insoluble under acidic conditions in water but soluble in alkaline conditions, hence contributing to its structure and long-term carbon sequestration. In turn, FA consists of more chemically reactive, smaller, water-soluble molecules that contribute much to the nutrient exchange and short-term processes of carbon dynamics. It is also called passive fractions of soil organic pools. The mean residence time varies from 250 to 1500 years. HA and FA represent significant fractions of soil organic carbon, indicators of soil health, and nutrient availability for overall ecosystem functioning.

S. Kale et al., *Soil Carbon Fractionation*, Sustainability Sciences in Asia and Africa,
https://doi.org/10.1007/978-981-95-4068-6_11

11.2 Principle

Humic acid (HA) and fulvic acid (FA) are fractions of soil organic matter that differ in their solubility and molecular structure. They are extracted using a *base extraction* method (Stevenson 1994) with sodium hydroxide (NaOH) (0.5 M) at pH 13.0, followed by an acid treatment using 2 N HCl at pH 2.0.

11.3 Reactions

Carbon in the sample is oxidized as follows:

$$2H_2\,Cr_2O_7 + 3C + 6H_2SO_4 = Cr_2\left(SO_4\right)_3 + 3CO_2 + 8H_2O$$

Thus, $2H_2\,Cr_2O_7$ or $2K_2\,Cr_2O_7 = 3C$
Or 1 mL of 1 N $2K_2\,Cr_2O_7 = 0.003$ g of C

11.4 Reagents

1. Sodium hydroxide solution (0.5 N NaOH): 20 g of NaOH in distilled water and make the volume up to 1 L.
2. Sodium pyrophosphate decahydrate solution (0.1 M): 44.606 g of $Na_4P_2O_7{\cdot}10H_2O$ in 1 L.
3. Hydrochloric acid solution (2 N HCl, pH 2).
4. Hydrochloric acid-hydrofluoric acid solution (HCl-HF): 5 mL of concentrated HCl and 5 mL of 52% HF in 990 mL of distilled water.
5. Use Dowex-50 cation exchange resin.

11.5 Procedure

A. *Estimation of HA and FA*

1. Place 12 g 2 mm size soil on filter paper of conical size placed on funnel, which is kept on conical flask.
2. Add 2 N HCL drop by drop to remove calcium carbonate till effervesces from the soil are stopped.
3. Wash the soil (2–3 times) with D.W.
4. After washing, keep the soil in the oven at 400–500 °C temperature till it dries.
5. Weigh 10 g of dried soil in a 500 mL conical flask.

6. Add 200 mL 0.5 N NaOH.
7. Shake on a shaker for 12 hours (Brown color will be developed).
8. Centrifuge at 6000–7000 rpm for 20–30 minutes.
9. Collect upper brown color supernatant in 500 mL conical flask.
10. Repeat washing steps with 0.5 N NaOH for 2–3 times.
11. Collect all the supernatant together from each washing in the 500 mL conical flask (obtained dissolved HA and FA in alkali).
12. Adjust the pH of supernatant collected in a conical flask up to 1.5–1.8 with 2 N HCL.
13. Kept conical flask as it is for 16 hours without disturbing (there will be precipitation of HA with acid).
14. Centrifuge at 7000–8000 rpm for 20 minutes.
15. Collect all the supernatant, i.e., fulvic acid (yellow colored).
16. Remaining precipitate material in the tube, i.e., humic acid (brown to dark colored).
17. Wash the humic acid with HCL: HF mixture (3:2) to remove cations and anions from the humic acid.
18. Repeat washing (3–4 times) and keep it in the oven at 40–50 °C till it dry.
19. Pass the separated fulvic acid through a sincere glass funnel to remove cations and anions by using Dowex 60 cations and anions exchanger resins.
20. A light-yellow color is developed (after the removal of cations and anions) (Flowchart 11.1 and Fig. 11.1).

B. *Determination of HA-C and FA-C by Wet Digestion Method*

Use separate humic acid and fulvic acid to estimate organic carbon by adopting the method of Walkley and Black.

Flowchart 11.1 Humic acid (HA) and fulvic acid (FA)

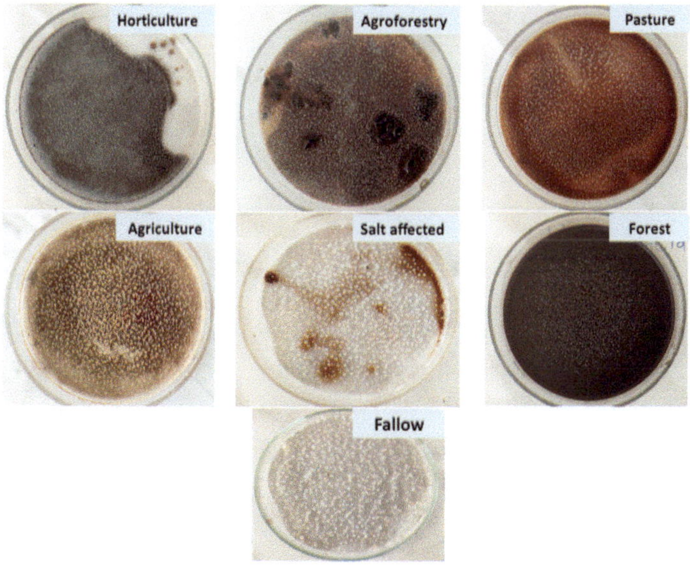

Fig. 11.1 Overview of laboratory analysis of humic acid carbon fractions from different land use systems

11.6 Calculation

$$\mathrm{SOC}(\%) = \left(\frac{B - S}{B}\right) \times \frac{V \times N \times 0.003 \times 100}{\mathrm{Wt.of\ soil\,sample}(\mathrm{g})}$$

where,

B = Volmume of FAS required for titration to blank sample
S = Volume of FAS required for titration to soil sample
V = Volume of $K_2Cr_2O_7$ taken
N = Normality of $K_2Cr_2O_7$

Suggested Reading

Stevenson FJ (1994) Humus chemistry, genesis, composition, reactions, 2nd edn. Wiley, New York, p 196

Determination of Soil Organic Carbon by Wet Digestion Method

12

Abstract

Soil organic carbon (SOC), a key component of soil organic matter that supports fertility and structure, is commonly estimated through chromic acid oxidation followed by redox titration using ferrous salts. This classical method, first introduced by Schollenberger in 1927, remains widely used for assessing soil carbon content and sustainability.

Keywords

Soil organic carbon · Walkley and Black method · Wet digestion · Potassium dichromate oxidation · Redox titration · Soil fertility indicator · Soil carbon assessment

12.1 Introduction

The important constituents of soil organic matter are soil organic carbon (SOC), comprising decomposed plant and animal residues, soil microbes, and root exudates. It is vital in maintaining soil health, fertility, and structure. SOC acts as a nutrient buffer while enhancing water retention and contributes to good aeration in the soil.

12.2 Principle

Soil organic matter undergoes oxidation with chromic acid, and the remaining potassium dichromate is back titrated using ferrous ammonium sulfate or ferrous sulfate (redox titration).

The idea that soil organic matter could be oxidized by applying a heated solution of $K_2Cr_2O_7$ and H_2SO_4 was initially put out by Schollenberger in 1927.

S. Kale et al., *Soil Carbon Fractionation*, Sustainability Sciences in Asia and Africa,
https://doi.org/10.1007/978-981-95-4068-6_12

$$2Cr_2O_7^{2-} + 3C^{o} + 16H^{+} \rightarrow 4Cr^{3+} + 3CO_2 + 8H_2O$$

The excess $Cr_2O_7^{2-}$ is titrated with ammonium ferrous sulfate [$Fe(NH_4)_2(SO_4)_2 \cdot 6H_2O$] following the reaction, and it is assumed that the $Cr_2O_7^{2-}$ that was reduced during the reaction with soil is equal to the organic carbon that is in the sample.

12.3 Equipment and Reagents

- 250-mL Erlenmeyer flasks.
- Burette (digital).

Potassium dichromate ($K_2Cr_2O_7$), 1 N (0.167 M): To prepare this solution, dissolve 49.04 g of reagent grade $K_2Cr_2O_7$ (previously dried at 105 °C) in distilled water and then dilute the solution to a final volume of 1000 mL.

Sulfuric acid (H_2SO_4): Use concentrated sulfuric acid with a minimum concentration of 96%. If there is chloride (Cl) present in the soil, add silver sulfate (Ag_2SO_4) to the acid at a rate of 15 g per liter.

Phosphoric acid (H_3PO_4): Use concentrated phosphoric acid.

1,10-Phenanthroline-ferrous complex, 0.025 M: To prepare this complex, dissolve 14.85 g of 1,10-phenanthroline monohydrate and 6.95 g of ferrous sulfate heptahydrate ($FeSO_4 \cdot 7H_2O$) in water. Then, dilute the solution to a final volume of 1000 mL.

Barium diphenylamine sulfonate: Prepare a 0.16% aqueous solution.

Ferrous sulfate heptahydrate ($FeSO_4 \cdot 7H_2O$) solution, 0.5 N: To prepare this solution, dissolve 140 g of reagent grade $FeSO_4 \cdot 7H_2O$ in water. Add 15 mL of concentrated H_2SO_4.

After cooling the solution, dilute it to a final volume of 1000 mL. Standardize this reagent daily by titrating it against 10 mL of 1 N $K_2Cr_2O_7$ as described below.

12.4 Procedure

1. Weigh 1 g of finely ground soil, sieved through a 0.5 mm mesh, and transfer it into a 500-mL Erlenmeyer flask.
2. Add 10 mL of 1 N potassium dichromate ($K_2Cr_2O_7$) to the flask, and gently swirl to mix the soil into the solution.
3. Quickly add 20 mL of concentrated sulfuric acid (H_2SO_4), pouring it directly into the soil mixture. Gently swirl the flask to combine the soil and chemicals, then swirl more vigorously for 1 min to ensure thorough mixing.
4. Allow the flask on an asbestos sheet for approximately 30 min.
5. Pour 200 mL of deionized water into the flask.
6. Add three to four drops of 1,10-phenanthroline indicator, then titrate the mixture using 0.5 N ferrous sulfate ($FeSO_4$). Near the endpoint, the solution will turn from greenish to dark green. At this point, continue adding ferrous sulfate drop

by drop until the color shifts sharply from blue to red (maroon), when observed against a white background.

7. To standardize the potassium dichromate solution, perform a blank titration following the same procedure but without the soil sample. Repeat the process with less soil if more than 75% of the dichromate is reduced (Figs. 12.1 and 12.2; Flowchart 12.1).

Fig. 12.1 Overview of laboratory analysis of soil organic carbon fractions (humic and fulvic acid carbon)

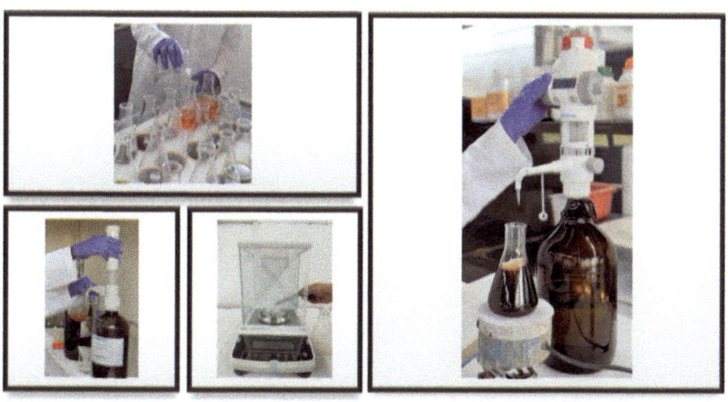

Fig. 12.2 Overview of laboratory analysis of soil organic carbon. (The Charles Renard Analytical Laboratory (CRAL), ICRISAT)

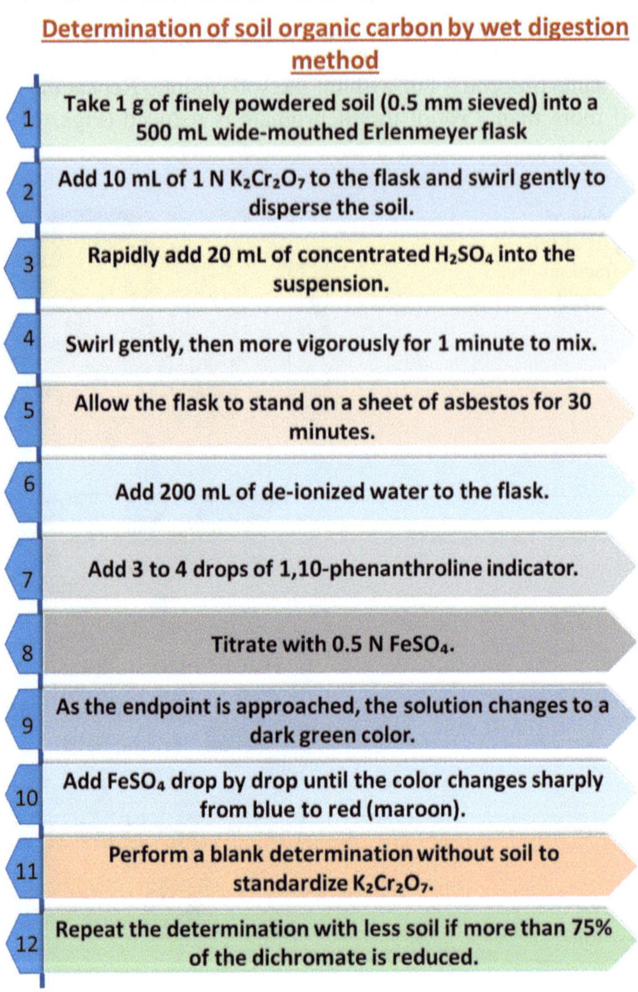

Flowchart 12.1 Determination of soil organic carbon by wet digestion method

12.5 Calculation

Compute the outcomes utilizing the subsequent formula, applying a correction factor of f = 1.30 or an alternative, more appropriate value determined through experimentation:

$$\text{Organic C}(\%) = \frac{(\text{mL Blank} - \text{mL Sample}) \times (\text{Normality of FeSO}_4) \times (0.003) \times (100)}{\text{Weight of water} - \text{free soil}(\text{g})}$$

12.6 Critical Limit

Soil	% of OC
Very low	0.25
Low	0.25–0.5
Medium	0.5–0.75
High	0.75–1.0
Very high	>1.0

Suggested Reading

Charles MJ, Simmons MS (1986) Methods for the determination of carbon in soils and sediments: a review. Analyst 111:385–390

Mebius LJ (1960) A rapid method for the determination of organic carbon in soil. Anal Chim Acta 22:120–124

Nelson DW, Sommers LE (1982) Total carbon, organic carbon, and organic matter. In: Page AL (ed) Methods of soil analysis, part II. Am. Soc. Agron. Inc and Soil Sci. Soc. Am. Inc, Madison, WI, pp 539–579

Nelson DW, Sommers LE (1996) Total carbon, organic carbon, and organic matter. In: Page AL et al (eds) Methods of soil analysis, part 2, Agronomy, vol 9, 2nd edn. Am. Soc. of Agron., Inc, Madison, WI, pp 961–1010

Walkley A, Black IA (1934) An examination of the Degtjareff method for determining organic carbon in soils: effect of variations in digestion conditions. Soil Sci 63:251–263

13 Determination of Total Soil Organic Carbon (Dry Combustion Method)

Abstract

The Walkley–Black method, a widely used technique for estimating soil organic carbon, offers rapid and cost-effective analysis despite incomplete oxidation, typically recovering 60–86% of SOC and requiring a correction factor for accuracy. Despite these drawbacks, the method continues to be a standard in soil testing laboratories for rapid SOC assessment. The dry combustion method, conducted at high temperatures, determines the total carbon content of soils using a CHN analyzer.

Keywords

Total soil organic carbon · Dry combustion method · CHN analyzer · CO2 detection · Non-dispersive infrared detector (NDIR) · High-temperature oxidation · Soil carbon quantification

13.1 Introduction

The Walkley-Black technique is widely used in soil testing laboratories to assess soil organic carbon. Nevertheless, the organic carbon in the soil is frequently not completely oxidized using this process. Research indicates that this process usually recovers between 60 and 86% of the organic carbon, with an average recovery of 76% (Walkley and Black 1934). It is standard procedure to apply a correction factor of 1.33 to the results in order to account for the partial oxidation and the absence of site-specific correction variables. The Walkley-Black technique is still widely used because of its ease of use, quickness, and low equipment needs, even with its drawbacks (Nelson and Sommers 1996).

S. Kale et al., *Soil Carbon Fractionation*, Sustainability Sciences in Asia and Africa,
https://doi.org/10.1007/978-981-95-4068-6_13

The quantity of organic carbon in a sample can be determined directly or calculated by subtracting the inorganic carbon content from the total carbon. In soils containing no inorganic carbon, the total carbon is equivalent to the organic carbon. To check for the presence of carbonates, a simple test can be conducted by adding a few drops of HCl (1–4 N HCl is often used) and observing if the sample bubbles. Two common acids, HCl and H2SO4, are typically used to remove carbonates until the bubbling stops. Numerous methods exist to determine organic carbon in soils. Two approaches are used for the determination of carbon.

1. Dry combustion method
2. Wet oxidation method

The materials are broken down in an acidic mixture using the wet oxidation method, and the amount of CO2 that is released is captured in NaOH.

13.2 Principle

The total carbon content of soils is determined by dry combustion at high temperatures in a furnace, which collects and detects evolved CO2 (Tiessen and Moir 1993). Dry combustion techniques involve sample combustion and quantification.

13.3 Equipment

Total organic carbon analyzer/CHN analyzer

13.4 Procedure

In the dry combustion technique, samples smaller than 0.2 mm are placed into an inert boat weighing about 250 mg, which is then inserted into a furnace heated to a temperature of 900–1000°C. The sample is burned at high temperatures in the presence of pure oxygen. The resulting CO2 gas is detected using a non-dispersive infrared detector (NDIR), and the carbon content of the sample is automatically displayed as the instrument is pre-calibrated for total organic carbon using a standard like sucrose. Analyses are halted once a known volume of gas has been collected or after a set amount of time has passed, and the total organic carbon content is calculated. The percentage of carbon should be reported on an oven-dry basis, so adjustments for moisture content will be necessary if the samples are only air-dried before analysis (Flowchart 13.1).

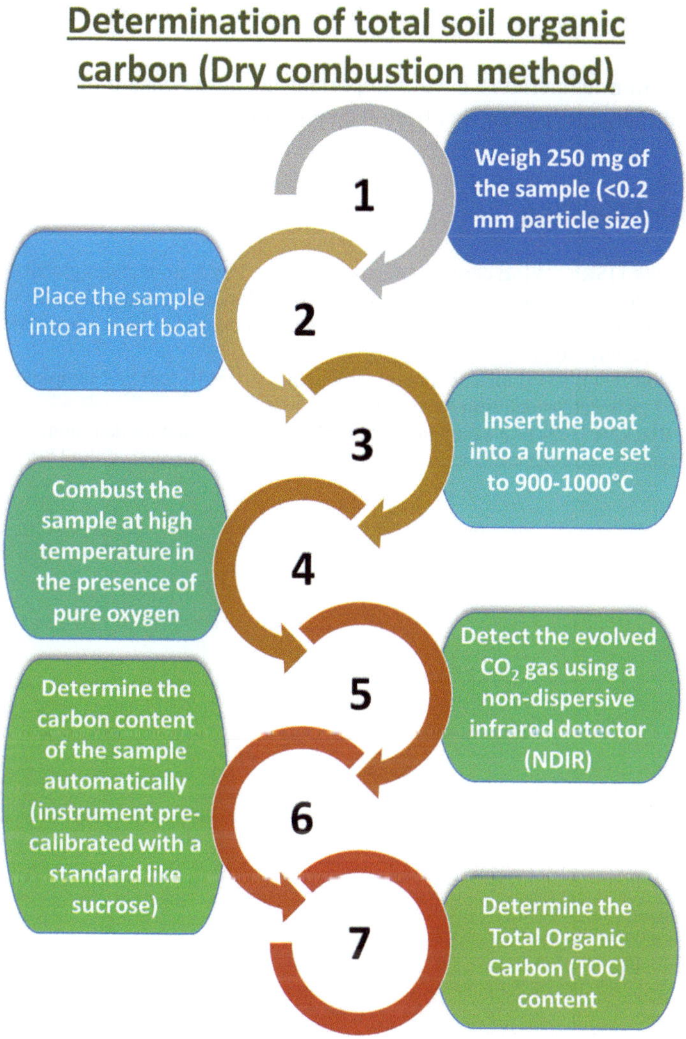

Flowchart 13.1 Determination of total soil organic carbon (dry combustion method)

13.5 Note

The combustion temperature is very important. It should ensure the elimination of all types of inorganic carbonates from the soil sample prior to TOC determination.

Suggested Reading

Batjes NH (2006) ISRIC-WISE derived soil properties on a 5 by 5 arc-minutes global grids, report 2006-02. ISRIC-World Soil Information, Wageningen. Available through: http://www.isric.org

Nelson DW, Sommers LE (1996) Total carbon, organic carbon, and organic matter. Methods Soil Anal Part 3 5:961–1010

Tiessen H, Moir JO (1993) Total and organic carbon. In: Carter ME (ed) Soil sampling methods of analysis. Lewis Publishers, Ann Arbor, MI, pp 187–211

Walkley A, Black IA (1934) An examination of the Degtjareff method for determining soil organic matter, and a proposed modification of the chromic acid titration method. Soil Sci 37(1):29–38

Total Organic Carbon in Soil Extracts: Titration Method

14

Abstract

Total Organic Carbon (TOC), a key indicator of soil health and carbon sequestration, is commonly determined by wet oxidation using potassium dichromate and sulfuric acid, followed by titration with ferrous ammonium sulfate for accurate quantification. This approach remains widely used due to its simplicity, cost-effectiveness, and suitability for routine laboratory analysis. Accurate TOC measurement provides essential insights into soil quality and sustainability.

Keywords

Total organic carbon in soil extracts · Titration method · Potassium dichromate oxidation · Ferrous ammonium sulfate · Microbial biomass carbon · Water-soluble organic carbon · Low-carbon soils

14.1 Introduction

Total Organic Carbon (TOC) in soil is a key indicator of soil health, fertility, and carbon sequestration potential. It reflects the amount of carbon bound in organic compounds, which directly influences nutrient cycling, microbial activity, and overall soil structure. The titration method for TOC determination commonly based on wet oxidation using potassium dichromate ($K_2Cr_2O_7$) and sulfuric acid (H_2SO_4) provides a reliable and widely used approach for quantifying oxidizable organic carbon in soil extracts. During this process, organic matter is oxidized, commonly based on wet oxidation using $K_2Cr_2O_7$ and H_2SO_4, and the unreacted dichromate is subsequently titrated with a standardized ferrous ammonium sulfate (FAS) solution to determine the amount of carbon oxidized. This method remains valuable in soil science due to its simplicity, cost-effectiveness, and suitability for routine laboratory analysis.

S. Kale et al., *Soil Carbon Fractionation*, Sustainability Sciences in Asia and Africa,
https://doi.org/10.1007/978-981-95-4068-6_14

14.2 Principle

Total organic carbon (TOC) in soil extracts is determined by wet oxidation using potassium dichromate in a strongly acidic medium, followed by back-titration with ferrous ammonium sulfate (FAS). Organic carbon present in the soil extract is oxidized by a known excess of potassium dichromate during digestion with concentrated sulfuric acid. The amount of dichromate reduced is proportional to the organic carbon content of the sample. The unreduced excess dichromate is then titrated with standardized acidified ferrous ammonium sulfate using o-phenanthroline as an indicator. The difference in titrant volume between blanks and samples is used to calculate organic carbon concentration.

14.3 Reagents

1. Ferrous ammonium sulfate hexahydrate.
2. Potassium diehromate, 0.0667 M
3. Sulfuric acid, conc.
4. Ferrous Ammonium Sulfate, Acidified, 0.033 M: Dissolve approximately 12.940 g of ferrous ammonium sulfate hexahydrate in 900 mL of water. Sulfuric acid (50 mL) should then be added. After letting cool, top off with 1000 mL of water and stir.
5. Solution for indicator: In 100 mL of water, dissolve 0.695 g of ferrous ammonium sulfate hexahydrate and 1.485 g of o-phenantholine monohydrate (WARNING: this is poisonous)

14.4 Procedure

Fill a digestion tube with 4.00 mL of the extract sample. Add 1 mL of potassium dichromate (0.0667 M). Stir constantly as you add 5 mL of pure sulfuric acid. Prepare two blank tubes (i.e., two tubes with reagents but no extracts). After heating the sample tubes and one blank to 150 °C for 30 min, remove and let them cool. Keep the other blank cold. Measure and transfer the contents of the tubes into Titrate test samples and blanks with acidified ferrous ammonium sulfate solution, ensuring thorough mixing with a magnetic stirrer. The endpoint is reached when the color changes from green/violet to red. Note the titers for the heated blank (mL-HB), the unheated blank (mL-UB), and each sample (mL-Sample).

Note: Due to oxidation, the acidified ferrous ammonium sulfate solution needs to be standardized every day. This will determine M, its molarity:

1. Pipette 1 mL potassium dichromate (0.0667 M) into a conical flask.
2. Add the indicator solution in 0.3 mL (3–4 drops). (A mouthpipette should not be used.)
3. Titrate ferrous ammonium sulfate solution that has been acidified, and note how many milliliters (T) were used.

14.5 Calculation

$$OC(\%) = \{(A^{*} M^{*} 0.003) / g\}^{*} (E / S)^{*} 100$$

where,

T = Standardization titer
M = Molarity of FAS (Approx. 0.033 m)
A = (mL HB – mL sample) * {(mLUB – mL HB)/mL UB} + (mL HB – mL sample)
g = Dry soil mass
E = Extraction Volume (mL)
S = Digest sample Volume (mL)
$M = 0.4/T$

Suggested Reading

Nelson DW, Sommers LB (1982) Total carbon, organic carbon and organic matter. In: Page AL, Miller RH, Keeney DR (eds) Methods of soil analysis. Part 2. American Society of Agronomy, Madison, pp 539–579

Estimation of Total Carbon and Total Nitrogen in Soil Samples by Dumas Method

15

Abstract

The Dumas method, a rapid and precise technique for determining total carbon and nitrogen in soils, involves high-temperature combustion under oxygen-rich conditions with detection of evolved gases using TC-TN analyzers. This method provides accurate measurements of TC and TN, making it a preferred approach for soil fertility and nutrient analysis.

Keywords

Total carbon · Total nitrogen · Dumas method · TC-TN analyzer · Dry combustion · Infrared detector · Thermal conductivity detector (TCD) · Soil nutrient analysis

15.1 Introduction

The Dumas method is a widely used technique for estimating total carbon (TC) and total nitrogen (TN) in soil samples. This process involves the combustion of a soil sample in a high-temperature furnace (typically 900–1000 °C) under an oxygen-rich environment. The carbon and nitrogen present in the sample are oxidized to CO_2 and NO_2, respectively.

Using a TC-TN analyzer, these gases are subsequently passed through a series of detectors (such as thermal conductivity or infrared detectors), which quantify the concentration of CO_2 for total carbon and the concentration of nitrogen gas for total nitrogen. This method is rapid, precise, and does not require prior sample preparation such as acid digestion.

S. Kale et al., *Soil Carbon Fractionation*, Sustainability Sciences in Asia and Africa,
https://doi.org/10.1007/978-981-95-4068-6_15

15.2 Principle

The TC-TN Analyzer (Primacs SNC 100) is used to determine the levels of carbon and nitrogen in soil by combusting the sample in pure oxygen at 1100 °C. During combustion, the gas mixture contains CO_2 and nitrogen oxides, which are carried by helium through a splitter. The CO2 is directed to an IR detector for measuring carbon, while the remaining gas mixture passes through a copper reduction oven to convert nitrogen oxides (mainly N2O) to N2 and then over CuO to convert CO to CO2. The carrier gas, containing the N2-CO2 mixture, is then passed through scrubbers to absorb the CO2 and H2O, and finally, the volume of N2 gas is measured using the TCD detector.

15.3 Apparatus

- TC-TN Analyzer (Model—Skalar Primacs SNC 100)
- 100 positions—Integrated auto sampler with a transparent lid
- Quartz/ceramic crucibles
- Tweezer
- Precision balance (accuracy of 0.0001 g)
- High-purity helium and oxygen cylinders

15.4 Methodology

- Check the pressures of helium and oxygen gases and turn on the instrument.
- Allow the instrument to stabilize for 30 min.
- Set up the method parameters for temperature and load the method to heat the combustion oven.
- In the auto sampler rack, start by placing five empty crucibles as blanks and five calibration standards (EDTA/Glycine) with different weights based on the sample range for C and N.
- Weigh approximately 200–300 mg of finely powdered soil samples, taking note of the accurate weights and using a tweezer to avoid contamination.
- Place the auto sampler rack in its position on the instrument. The sample rack is removable, and reusable quartz crucibles are used for samples weighing up to 3 g of solid material.
- Create a table by entering the software's blanks, standards, sample positions, and weights.
- Start the analysis once the temperature of the oven is stabilized.
- Introduce samples into the analyzer through a vertical sample introduction system.
- After the analysis, the sample ashes remain in the crucible and are removed, preventing ash build-up in the combustion zone and reducing maintenance.
- After completion, the results of each sample are displayed in a separate window in the software in percentage (Fig. 15.1).

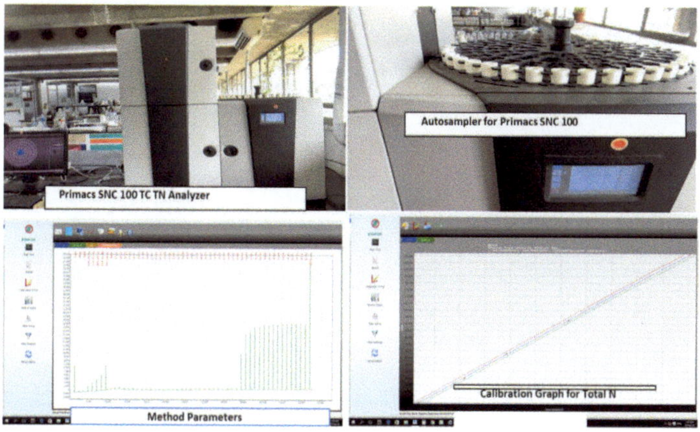

Fig. 15.1 Setup for estimation of total carbon and total nitrogen in soil by Dumas method

Soil Calcium Carbonate Equivalent by Volumetric Calcimeter Method

16

Abstract

Soil carbonates, primarily calcium carbonate and dolomite, are key regulators of soil pH and carbon cycling, commonly quantified by acid dissolution and CO_2 measurement or equivalent calcium carbonate determination. Understanding soil carbonate content is essential for evaluating soil chemistry, fertility, and its role in the global carbon cycle.

Keywords

Soil calcium carbonate equivalent · Volumetric calcimeter method · CO_2 evolution · Acid neutralization · Carbonate-rich soils · Calcium carbonate estimation

16.1 Introduction

Calcium carbonate (CaCO3), dolomite [CaMg(CO3)2], and magnesian calcite make up over 90% of natural carbonates and play a crucial role in the global carbon cycle. They are commonly found in arid, semi-arid, and sub-humid soils, where they influence soil properties such as pH and reactions like sorption-desorption and precipitation-dissolution.

Soil carbonates are measured by dissolving them in an acid solution and then evaluating H+ consumption, as well as the amounts of calcium (Ca) and magnesium (Mg) or the production of carbon dioxide (CO2). GLOSOLAN has standardized the methods for quantifying calcium carbonate content. In soils affected by carbonates, the calcium carbonate content can range from negligible to over 80%. Calcium carbonate equivalents are determined if the soil pH is above seven or if effervescence occurs after treatment with 1 M HCl. Carbonates are typically found in neutral to alkaline soils, but solid-phase carbonates, such as nodules, can also be present in some acidic environments. Soil carbonates are typically measured by dissolving

S. Kale et al., *Soil Carbon Fractionation*, Sustainability Sciences in Asia and Africa,
https://doi.org/10.1007/978-981-95-4068-6_16

carbonates in an acid solution and then determining either H^+ consumption, Ca and Mg production, or CO_2 production.

$$CaCO_3 + 2H^+ \rightarrow Ca2 + +CO_2 + H_2O \quad \text{and}$$

$$CaMg(CO_3)_2 + 4H^+ \rightarrow Ca2 + +Mg2 + +2CO_2 + 2H_2O$$

16.2 Scope and Field of Application

This method estimates soil carbonate percentage by digesting with excess acid and measuring emitted CO2 using a volumetric calcimeter. It can be used for all soils but may overestimate calcium carbonate equivalent (CCE) in high organic matter soils.

16.3 Principle

The process involves treating carbonates with hydrochloric acid to measure the volume of carbon dioxide released, indicating soil carbonate decomposition. Various calcimeter devices determine calcium carbonate equivalent (CCE) volumetrically. When organic matter decomposition is absent, CO2 measurement provides a direct assessment of carbonates, but precautions are needed to avoid interference from organic matter oxidation. Manganese dioxide can disrupt acid dissolution, and using salted water or phosphoric acid minimizes CO2 solvation. Adding octyl alcohol also helps reduce foaming when treating high carbonate soils with HCl.

16.4 Apparatus

1. Volumetric calcimeter
2. Magnetic stirrer
3. Electronic balance
4. Volumetric dispenser

16.5 Reagents and Standards

- *Preparation of 4 N Hydrochloric Acid (HCl) Solution*:
  - To prepare a 4 N HCl solution, carefully measure 393.3 mL of concentrated HCl (32% concentration, specific gravity of 1.16) and add it slowly to 400 mL of deionized water.
  - Once the acid is mixed with water, top up the solution with additional deionized water until the total volume reaches 1 L.

- *Preparation of 10% Sodium Chloride (NaCl) Solution*:
  - To make a 10% NaCl solution, weigh 100 g of sodium chloride.
  - Transfer the NaCl to a 1-L flask and add enough deionized water to reach a total volume of 1 L, ensuring thorough dissolution.
- *Reagent-Grade Calcium Carbonate (CaCO3)*:

16.6 Sample Preparation

- The soil samples should be air dried
- Once dried, crush the soil and pass it through a 10-mesh sieve with a 2 mm pore size to remove larger particles.
- To achieve a uniform sample, take the sieved soil and grind it further into a fine powder using a suitable mortar and pestle, ensuring the sample is well homogenized and ready for analysis.

16.7 Procedure

16.7.1 Calibration of Calcimeter

1. Accurately weigh fine-grained, reagent-grade calcium carbonate (CaCO3) to the nearest 0.1 mg and place it in reaction flask A. Recommended sample sizes are approximately 0.1000 g, 0.2000 g, 0.3000 g, 0.4000 g, and 0.6000 g. Add 20 mL of deionized water to the flask and include a stirring bar.
2. Measure 7 mL of the acid solution and pour it into container B.
3. Set up the reaction flask in the system. Open the three-way stopcock (C) to the air and adjust the liquid level in the measuring burette (D) to precisely 0 mL by raising or lowering the leveling bulb (E).
4. Seal the system from the atmosphere by rotating the stopcock (C) by 180°. Lower the leveling bulb by approximately 2 cm.
5. At the same time, begin adding the HCl solution from container B into the sample flask by tilting it and lowering the leveling bulb. Ensure that the liquid level in the bulb remains 1–2 cm below the level in the burette (D).
6. Switch on the magnetic stirrer at a slow speed. If no stirrer is available, manually shake the flask continuously.
7. Once the liquid level in the gas burette stabilizes, equalize the liquid levels in both the leveling bulb (E) and the burette (D). Record the volume of CO2 that is released (Flowchart 16.1).

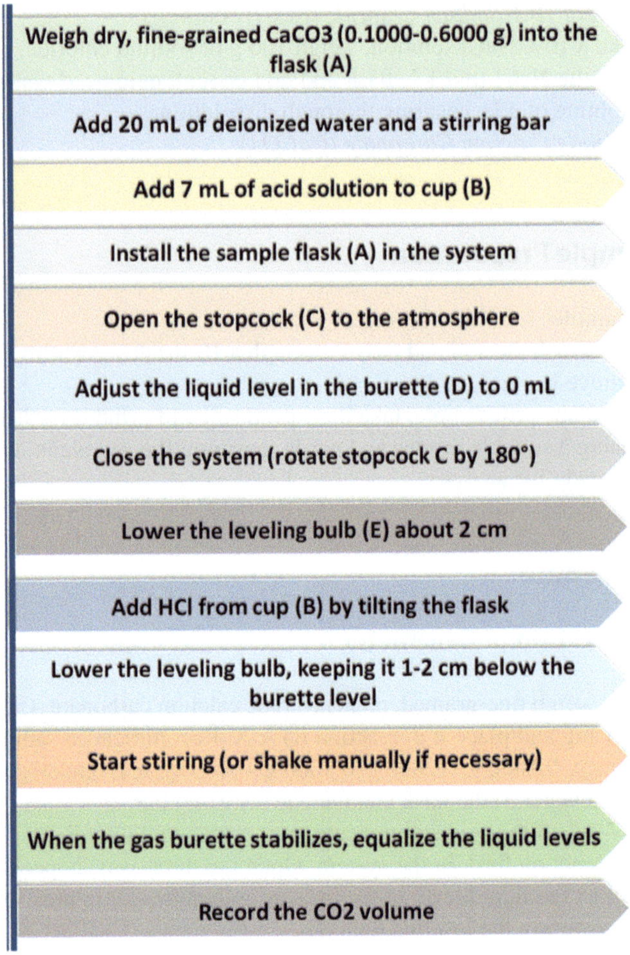

Flowchart 16.1 Determination of soil calcium carbonate equivalent by volumetric calcimeter method

16.7.2 Determination of the Total Carbonate in Soils

Add 0.5 to 5.0 g ± 1 mg of soil sample to the reaction flask.

16.8 Sample Weight Estimation

Effervescence and appearance after adding HCl (1 M)	$CaCO_3$ expected (%)	Sample weight (g)
None, very slight, slight but for a short time	<8	5.0
Strong, bubbles form, low foam	8–16	2.5
Violent, thick foam forms quickly	>16	1.0

16.9 Calculations

16.9.1 Calibration of Calcimeter

Correct the CO_2 volume for the standards and samples by subtracting the CO_2 volumes for the blank as follows:

$$\text{VCO2}(\text{corr}) = \text{VCO2} - \text{VCO2}(\text{blank})$$

Plotting calibration curve: Plot actual CaCO3(WCaCO3) on the *y*-axis vs. VCO2(corr) on the *x*-axis. Then, use linear regression to relate the corrected volume of CO2 released to the mass of pure CaCO3 and, thus, derive the regression equation.

16.9.2 Determination of Carbonate In Soils

Calcium carbonate equiv. is calculated as follows (Fig. 16.1):

$$\text{CaCO}_3\ \text{equiv.,\%} = \left(W_{\text{CaCO3}} / W_{\text{soil}}\right) \times 100$$

Where: W_{CaCO3} = Weight of $CaCO_3$ calculated from the calibration curve (g)
W_{soil} = Weight of soil (g)

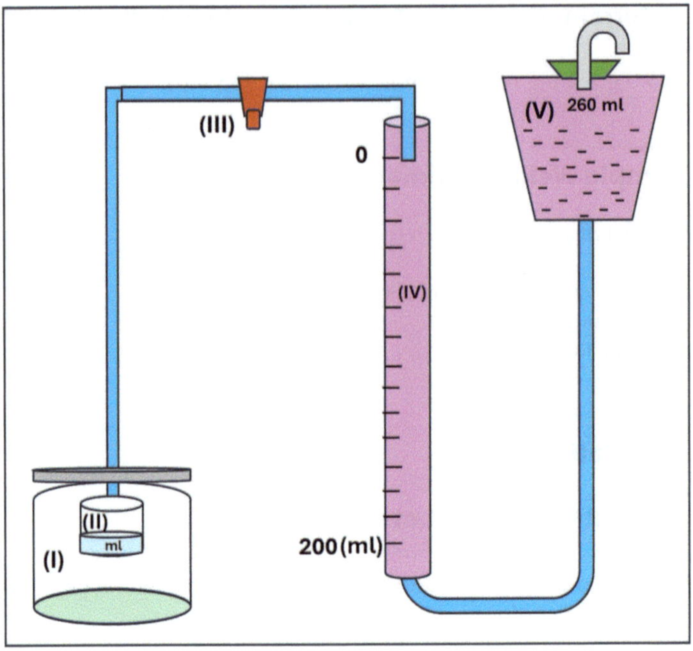

Fig. 16.1 Calcimeter apparatus: (I) reaction flask, (II) cup, (III) stopcock, (IV) burette, (V) leveling bulb

Suggested Reading

Allison L (1960) Wet-combustion apparatus and procedure for organic and inorganic carbon in soil. Soil Sci Soc Am J 24:36–40

AOAC (1998) Peer verified methods program. Manual on policies and procedures. AOAC International, Gathersburg, MD

Bundy L, Bremner J (1972) A simple titrimetric method for determination of inorganic carbon in soils. Soil Sci Soc Am J 36:273–275

FAO (2020) Standard operating procedure for soil calcium carbonate equivalent. Volumetric calcimeter method. FAO, Rome

Inskeep WP, Bloom PR (1987) Soil chemical factors associated with soybean chlorosis in Calciaquolls of western Minnesota. Agron J 79:779–786

Lal R (2006) Encyclopedia of soil science. CRC Press, Boca Raton, p 1923

Lal R, Kimble J (2000) Inorganic carbon and the global C cycle: research and development priorities. In: Global climate change and pedogenic carbonates. Lewis, Boca Raton, pp 291–302

Lal R, Kimble JM, Stewart B, Eswaran H (1999) Global climate change and pedogenic carbonates. CRC Press, Boca Raton, p 325

Lindsay WL, Thorne D (1954) Bicarbonate ion and oxygen level as related to chlorosis. Soil Sci 77:271–280

Loeppert RH, Suarez DL (1996) Carbonate and gypsum. Publications from USDA-ARS, UNL Faculty 504

Soil Survey Staff (2014) Keys to soil taxonomy. United States Department of Agriculture, Natural Resources Conservation Service

Assay of Soil Dehydrogenase Activity: TTC Method

17

Abstract

Soil dehydrogenase activity (DHA), a key indicator of microbial oxidative processes and soil health, is commonly measured by the TTC method, where triphenyl tetrazolium chloride is reduced to triphenyl formazan and quantified spectrophotometrically. DHA measurement provides valuable insights into microbial oxidative metabolism and soil biological functioning.

Keywords

Soil dehydrogenase activity · TTC method · Microbial activity · Triphenyl formazan (TPF) · Spectrophotometric assay · Soil enzyme indicators · Soil biological activity

17.1 Introduction

Soil dehydrogenase activity (DHA) is a commonly used biochemical indicator of both soil microbial activity and the soil's overall health. Dehydrogenases are intracellular enzymes that occur in all living microbial cells and play a vital role in the biological oxidation of organic matter by transferring hydrogen from substrates to acceptors. Measurement of DHA gives an idea about the microbial oxidative activities in soils. Among the currently used methods for estimating the activity of soil dehydrogenases, the most commonly applied technique is the TTC method.

S. Kale et al., *Soil Carbon Fractionation*, Sustainability Sciences in Asia and Africa,
https://doi.org/10.1007/978-981-95-4068-6_17

17.2 Principle

Assays of dehydrogenase activity in soil by the TTC method are conducted such that the colorless substrate, triphenyl tetrazolium chloride (TTC), is reduced to a red-colored product, triphenyl formazan (TPF), by the presence of dehydrogenase enzymes in active microbial cells. The reaction proceeds as illustrated below:

Substrate Conversion TTC is a water-soluble compound that is an artificial electron acceptor. TTC is reduced to insoluble red-colored TPF by active dehydrogenase enzymes from viable microbial cells.

$$\text{TTC}(\text{colorless}) + 2\text{H}+ \rightarrow \text{TPF}(\text{red})$$

Extraction and Quantification After incubation, the soil sample is treated with an organic solvent like methanol or ethanol to extract the formed TPF. Then, it is measured spectrophotometrically at a particular wavelength (near 485 nm), the intensity of the red color being directly proportional to dehydrogenase activity. TPF that has been formed points to the general metabolic activity of soil microorganisms since dehydrogenases take part in a great number of cells' oxidation processes.

17.3 Apparatus

(a) Spectrophotometer.
(b) Test tubes, 16 by 150 mm.

17.4 Reagents

1. Calcium carbonate (CaCO3), reagent grade.
2. Triphenyl formazan (TPF) standard solution: 100 mg of TPF dissolve in about 80 mL of methanol and adjust the volume to 100 mL with methanol.
3. 2,3,5-Triphenyltetrazolium chloride (TTC), 3%: 3 g of TTC in about 80 mL of H2O and adjust the volume to 100 mL with H2O.
4. Methanol.

17.5 Procedure

1. Mix 20 g of air-dried soil (particle size <2 mm) with 0.2 g of $CaCO_3$. Then, place 6 g of this mixture into each of the three test tubes.

2. Add 1 mL of a 3% aqueous solution of TTC and 2.5 mL of distilled water to each tube.
3. After mixing, this amount should create a small amount of free liquid at the soil surface.
4. Stir the contents of each tube with a glass rod, seal the tube, and place it in an incubator at 37 °C. After 24 h, remove the stopper, add 10 mL of methanol, recap the tube, and shake for 1 min.
5. Remove the stopper, centrifuge at 10,000 rpm for 10 min, and filter the suspension through a glass funnel into a 100-mL volumetric flask using a No. 42 Whatman filter.
6. Rinse the tube with methanol, add more in 10-mL increments, centrifuge again, and transfer the suspension into the volumetric flask. Dilute to a final volume of 100 mL using methanol. Measure the intensity of the red color using a spectrophotometer at a wavelength of 485 nm.
7. Use methanol as a blank for calibration.
8. Determine the total TPF produced using a calibration graph derived from the standard curve (Flowchart 17.1; Fig. 17.1).

17.6 Standard Curve

To prepare a graph, start by diluting 10 mL of TPF standard solution with ethanol to a total volume of 199 mL (resulting in a concentration of 100 μg of TPF per mL.) Then, using a volumetric flask (50 mL), pipette 0, 0.5, 1.0, 2.0, 3.0, and 4.0 mL of the diluted TPF standard solution and bring each up to 50 mL with methanol. This will give you the following concentrations: 0, 5, 10, 20, 30, and 40 μg TPF per mL.

Fig. 17.1 Assay of soil dehydrogenase activity

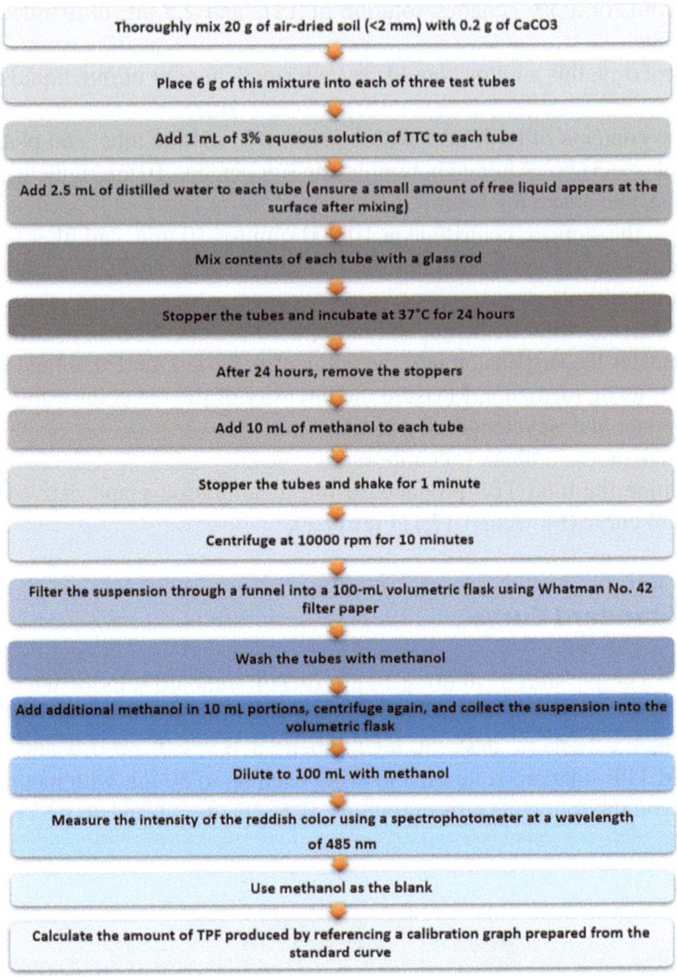

Flowchart 17.1 Assay of soil dehydrogenase activity

17.7 Calculation

Obtain the TPF concentrations (µg/mL) from the calibration curve, adjust for the control value, and compute accordingly:

$$\text{Dehydrogenase activity} = \text{O.D.value} \times \text{Dilution factor} \left(\mu g\,\text{TPF}\,\text{g}^{-1}\,24\text{h}^{-1}\right)$$

$$= \text{O.D.value} \times \frac{\text{Final vol. of extract}}{\text{Wt. of sample}}$$

The dehydrogenases that are measured are activities within the cells and not outside the cells. The procedure can be altered by using 10 g of soil, 10 mL of Tris buffer, and incubating at 27 °C for 15 h. Both organic and inorganic colloids can absorb TTC; therefore, it is important to optimize the TTC concentrations. TTC and TPF are sensitive to light, so making measurements in the dark is recommended. It is crucial to use only clear filtrate to measure the optical density. When measuring microbial activity through TTC reduction, removing plant roots and visible animals from soil samples is advisable. Dehydrogenase activities are very low in acidic soils (pH < 5). O2, NO^{3-}, Fe^{3+}, and $NO2^{-}$ also inhibit dehydrogenase activities, while inorganic P, Fe^{2+}, $SO4^{-}$, Cl, and Mn stimulate TTC reduction (Bremner and Tabatabai 1973). TPF is insoluble in water and stays at the reaction sites, making it useful for locating the reaction site. Pollutants affect TTC reduction to varying degrees, and TTC reduction does not always correlate with time.

17.8 Example Calculation

Step	Details
Spectrophotometer reading (O.D.)	O.D. value obtained from spectrophotometer
Using the standard curve (TPF concentration)	TPF concentration (e.g., 20 µg/mL from standard curve)
The final volume of the extract	100 mL
Weight of soil sample	6 g
Dehydrogenase activity (µg TPF/g soil/24 h)	Dehydrogenase activity = (O.D. value × 100 mL) / 6 g

Suggested Reading

Casida LE, Klein D, Santora R (1964) Soil dehydrogenase activity. Soil Sci 98:371–378

Bremner JM, Tabatabai MA (1973) Nitrogen-15 enrichment of soils and soil-derived nitrate. American Society of Agronomy, Crop Science Society of America, and Soil Science Society of America. J Environ Qual 2(3):363–365

Assay of Soil β-Glucosidase Activity

18

Abstract

Beta-glucosidase, a key enzyme in cellulose degradation and carbon cycling, serves as an important biochemical indicator of soil quality and organic matter turnover by catalyzing the hydrolysis of beta-glucosides into glucose. Measuring beta-glucosidase activity provides insights into carbon turnover and the soil's capacity to sustain biological processes essential for ecosystem functioning. This enzyme-based assessment is widely used to evaluate soil health and organic matter dynamics.

Keywords

β-glucosidase activity · p-Nitrophenyl glucoside (PNG) · Soil carbon cycle · Soil enzymes · Microbial activity · p-Nitrophenol (PNP) · Soil organic matter turnover · Enzymatic soil health indicators

18.1 Introduction

Beta-glucosidase is a vital enzyme in the soil ecosystem's carbon cycle. It plays a crucial role in breaking down cellulose, a major component of plant biomass, by catalyzing the hydrolysis of cellobiose and other beta-glucosides into glucose. The activities of beta-glucosidases are often used as a biochemical index of soil quality and organic matter turnover, reflecting microbial activity associated with carbon cycling. Measuring the activity of beta-glucosidase provides valuable information about the soil's capability to decompose organic matter and facilitate nutrient cycling, which is crucial for assessing soil health and fertility. The general equation for the reaction is glycosides + H_2O → sugar + aglycons.

S. Kale et al., *Soil Carbon Fractionation*, Sustainability Sciences in Asia and Africa,
https://doi.org/10.1007/978-981-95-4068-6_18

18.2 Apparatus

1. Spectro-photometer, Incubation flasks, stoppers, incubator.

18.3 Reagents

1. *Preparation of 0.05 M p-Nitrophenyl-β-D-glucoside (PNG) Solution*:
2. Dissolve 0.654 g of PNG in 40 mL of MUB (pH 6.0) and adjust to 50 mL. Refrigerate.
3. *Preparation of Modified Universal Buffer (MUB) Stock Solution*:
4. Prepare MUB stock by dissolving 12.1 g THAM, 11.6 g maleic acid, 14.0 g citric acid, and 6.3 g boric acid in 488 mL of 1 N NaOH. Dilute to 1 L and refrigerate.
5. *Preparation of Modified Universal Buffer (MUB) at pH 6.0*:
6. Adjust 200 mL of MUB stock to pH 6.0 with 0.1 M HCl. Dilute to 1 L.
7. *Preparation of 0.5 M Calcium Chloride ($CaCl_2$) Solution*:
8. Dissolve 73.5 g of $CaCl_2 \cdot 2H_2O$ in 700 mL water and dilute to 1 L.
9. *Preparation of Tris(hydroxymethyl)aminomethane (THAM) Buffers at pH 12 and 10*:
10. Dissolve 12.2 g THAM in 700 mL water, adjust pH to 12 or 10 with 0.1 M NaOH, and dilute to 1 L.
11. *Preparation of p-Nitrophenol Solution*:
12. Dissolve 1 g of p-nitrophenol in 700 mL water and dilute to 1 L. Refrigerate.

18.4 Procedure

1. Place 1 g of soil into a 50-mL Erlenmeyer flask.
2. Add 0.25 mL of toluene, 4 mL of MUB (pH 6.0), and 1 mL of PNG solution to the flask. Shake the flask briefly to mix the contents.
3. Seal the flask and incubate it at 37 °C. After 1 h, remove the stopper and add 1 mL of 0.5 M $CaCl_2$ and 4 mL of 0.1 M THAM buffer (pH 12). Shake the flask, then filter the soil suspension using Whatman no. 2 filter paper.
4. Measure the yellow color intensity using a spectrophotometer at 420 nm. Use a calibration curve to determine the concentration of p-nitrophenol in the filtrate (Flowchart 18.1).

If the filtered solution's color intensity surpasses that of the highest p-nitrophenol standard, dilute the solution with 0.1 M THAM (pH 10) until the colorimeter reading falls within the calibration range. Air-drying field-moist soils significantly increases the activity ofα and β-glucosidases and α- and β-galactosidases. The

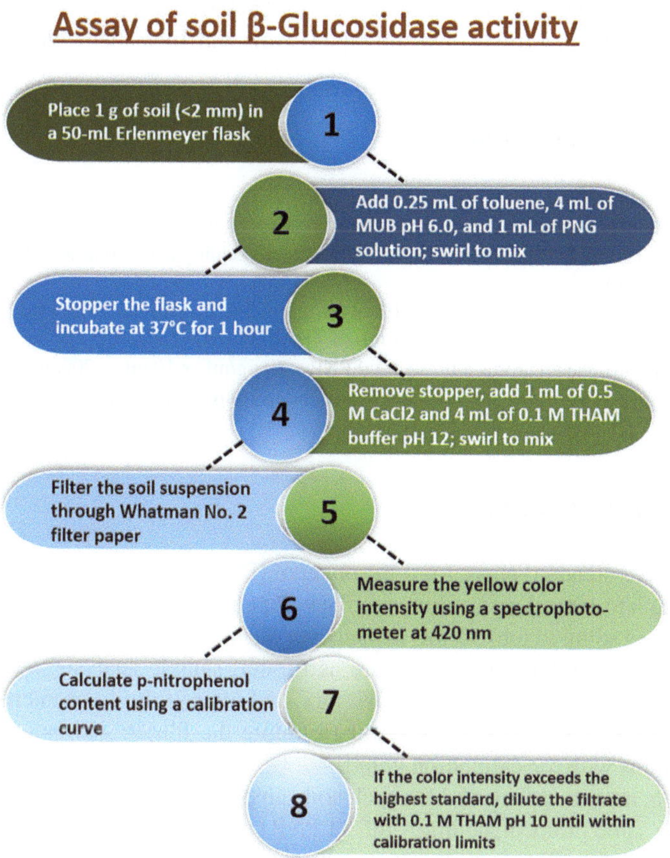

Flowchart 18.1 Assay of soil β-glucosidase activity

effects of trace elements and inorganic salts on enzyme activity vary depending on the specific element, enzyme, and soil type (Eivazi and Tabatabai 1990).

18.4.1 Example Calculation

Step	Details
Spectrophotometer reading	0.350 absorbance units at 420 nm
p-Nitrophenol concentration (from calibration curve)	20 μg/mL corresponding to 0.350 absorbance
Volume of soil extract	10 mL (4 mL of MUB, 1 mL of PNG, 1 mL of $CaCl_2$, 4 mL of THAM buffer)
Amount of p-nitrophenol produced	Total PNP = 20 μg/mL × 10 mL = 200 μg
β-glucosidase activity	β-glucosidase activity = 200 μg PNP/g soil/h

18.5 Standard Curve

To construct a standard curve for p-nitrophenol (PNP), dilute 1 mL of the standard PNP solution to 100 mL in a volumetric flask. Pipette 0, 1, 2, 3, 4, and 5 mL aliquots of this diluted solution into 50-mL Erlenmeyer flasks and adjust the volume to 5 mL with water. Follow the same procedure as for soil sample analysis: add 1 mL of 0.5 M $CaCl_2$ and 4 mL of 0.5 M NaOH, mix, and filter. If the color intensity of the filtrate exceeds that of the 50 μg standard, dilute the filtrate with water until the reading aligns with the calibration graph.

Controls must be performed for each soil sample to account for non-PNP color. Follow the same protocol, adding 1 mL of PNP solution after the $CaCl_2$ and NaOH addition but before filtration. Results can be expressed as μg PNP/g/2 h.

18.6 Calculation

$$\beta-\text{Glucosidase} = \left(\text{Sample O.D.} - \text{Blank O.D.}\right) \times \text{dilution factor}$$

$$\left(\mu g\,\text{PNG}\,g^{-1}\,\text{day}^{-1}\right) \times \text{graph factor}$$

$$= \left(\text{Sample O.D.} - \text{Blank O.D.}\right) \times \frac{\text{Vol.made}}{\text{Wt.of soil}} \times \text{graph factor}$$

Suggested Reading

Eivazi F, Tabatabai MA (1988) Glucosidases and galactosidases in soils. Soil Biol Biochem 20:601–606

Eivazi F, Tabatabai MA (1990) Factors affecting glucosidase and galactosidase activities in soils. Soil Biol Biochem 22:891–897

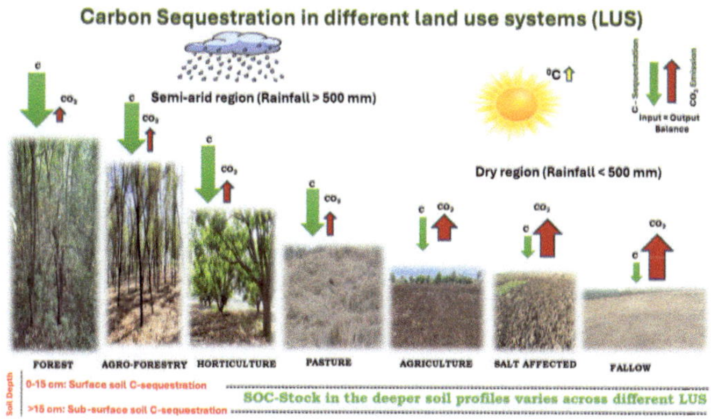

ICRISAT
INTERNATIONAL CROPS RESEARCH INSTITUTE FOR THE SEMI-ARID TROPICS

About

The International Crops Research Institute for the Semi-Arid Tropics (ICRISAT) is a pioneering International Organization committed to developing and improving dryland farming and agri-food systems to address the challenges of hunger, malnutrition, poverty, and environmental degradation affecting the 2.1 billion people residing in the drylands of Asia, Sub-Saharan Africa, and beyond.

ICRISAT was established under a Memorandum of Agreement between the Government of India and the CGIAR on the 28 March 1972. In accordance with the Headquarters Agreement, the Government of India has extended the status of a specified "International Organisation" to ICRISAT under section 3 of the United Nations (Privileges and Immunities) Act, 1947 of the Republic of India through Extraordinary Gazette Notification No. UI/222(66)/71, dated 28 October 1972, issued by the Ministry of External Affairs, Government of India.

Asia

ICRISAT - India (Headquarters)
Patancheru 502 324, Hyderabad
Telangana, India
Phone: +91 8455683071
Fax: +91 8455683074
Email: icrisat-ind@icrisat.org

ICRISAT - India (Liaison Office)
CG Centers Block
NASC ComplexDev Prakash Shastri Marg, New Delhi 110012, India
Phone: +91-11-25840294
Fax: +91 1125841294
Email: icrisat-ind@icrisat.org

West and Central Africa

ICRISAT - Mali
(Regional hub WCA)
BP 320 Bamako, Mali
Phone: +223 20 709200
Fax: 223 20 709201
Email: icrisat-mli@icrisat.org

ICRISAT - Niger
BP 12404
Niamey, Niger (via Paris)
Phone: +(227) 20722725, 20722626
Fax: +227 20734329
Email: icrisat-ner@icrisat.org

ICRISAT - Nigeria
PMB 3491
Sabo Bakin Zuwo Road
Tarauni, Kano, Nigeria
Phone: +234 7034889836
Email: icrisat-nga@icrisat.org

ICRISAT - Senegal
c/o Africa Rice
Mamelles Aviation, Villa 18
BP 24365 Dakar, Senegal
Phone: +221 338600706
Email: icrisat-sen@icrisat.org

Eastern and Southern Africa

ICRISAT - Kenya
(Regional hub ESA)
PO Box: 39063, Nairobi, Kenya
Phone: +254 20 7224550
Fax: +254 20 7224001
Email: icrisat-ken@icrisat.org

ICRISAT - Ethiopia
C/o ILRI Campus
PO Box 5689, Addis Ababa, Ethiopia
Phone: +251-11 617 2541
Fax: +251 11 646 1252, +251 11 646 4645
Email: icrisat-eth@icrisat.org

ICRISAT - Malawi
Chitedze Agricultural Research Station
PO Box 1096, Lilongwe, Malawi
Phone: +265 1 707 297/071/067/057
Fax: +265 1 707 298
Email: icrisat-mwi@icrisat.org

ICRISAT - Zimbabwe
Matopos Research Station
PO Box 776, Bulawayo, Zimbabwe
Phone: +263 292 809314/315
Fax: +263 383 307
Email: icrisat-zwe@icrisat.org

ICRISAT - Mozambique
(c/o IIAM) nr 2698 1st Floor, AV. FPLM
Maputo, Mozambique
Phone: +258 1 461657
Fax: +258 1 461581
Email: icrisat-moz@icrisat.org

ICRISAT - Tanzania
Plot 25, Mikocheni Light Industrial Area
Mwenge Coca-Cola Road, Mikocheni B,
PO Box 34441, Dar es Salaam, Tanzania
Email: icrisat-tza@icrisat.org

www.icrisat.org

MIX
Papier aus verantwortungsvollen Quellen
Paper from responsible sources
FSC® C105338

If you have any concerns about our products,
you can contact us on
ProductSafety@springernature.com

In case Publisher is established outside the EU,
the EU authorized representative is:
Springer Nature Customer Service Center GmbH
Europaplatz 3, 69115 Heidelberg, Germany

Printed by Libri Plureos GmbH
in Hamburg, Germany